计算机技术开发与应用丛书

OpenHarmony轻量系统
从入门到精通50例

戈 帅 ◎ 编著

清华大学出版社

北京

内 容 简 介

本书通过丰富的实战案例由表及里、深入浅出,以基础知识和案例相结合的方式,循序渐进地系统讲解 OpenHarmony 轻量系统应用开发技术。

全书共 6 章,第 1 章介绍了 OpenHarmony 轻量系统的背景、发展历程、环境搭建、工程创建与运行等情况。第 2~5 章通过 49 个案例详细介绍了 OpenHarmony 轻量系统的设备接口、系统接口、智能设备、网络通信等关键技术。第 6 章为综合实战案例——遥控小车,该综合实战案例从技术架构、功能实现、MQTT 通信协议等方面进行讲解,让读者掌握 OpenHarmony 轻量系统应用开发技术,提升读者综合开发能力。

本书主要面向对 OpenHarmony 轻量系统应用开发感兴趣的学生、开发人员或者相关从业人员,让更多的人掌握 OpenHarmony 轻量系统应用开发技术。

本书封面贴有清华大学出版社防伪标签,无标签者不得销售。
版权所有,侵权必究。举报: 010-62782989, beiqinquan@tup.tsinghua.edu.cn。

图书在版编目(CIP)数据

OpenHarmony 轻量系统从入门到精通 50 例/戈帅编著. —北京: 清华大学出版社,2023.10
(计算机技术开发与应用丛书)
ISBN 978-7-302-64229-9

Ⅰ. ①O… Ⅱ. ①戈… Ⅲ. ①移动终端－应用程序－程序设计 Ⅳ. ①TN929.53

中国国家版本馆 CIP 数据核字(2023)第 136003 号

责任编辑: 赵佳霓
封面设计: 吴 刚
责任校对: 时翠兰
责任印制: 丛怀宇

出版发行: 清华大学出版社
网　　址: https://www.tup.com.cn, https://www.wqxuetang.com
地　　址: 北京清华大学学研大厦 A 座　　　邮　　编: 100084
社 总 机: 010-83470000　　　邮　　购: 010-62786544
投稿与读者服务: 010-62776969, c-service@tup.tsinghua.edu.cn
质量反馈: 010-62772015, zhiliang@tup.tsinghua.edu.cn
课件下载: https://www.tup.com.cn, 010-83470236

印 装 者: 北京同文印刷有限责任公司
经　　销: 全国新华书店
开　　本: 186mm×240mm　　印　张: 17.75　　字　数: 398 千字
版　　次: 2023 年 12 月第 1 版　　印　次: 2023 年 12 月第 1 次印刷
印　　数: 1~2000
定　　价: 69.00 元

产品编号: 098782-01

前 言
PREFACE

OpenHarmony 操作系统是由开放原子开源基金会(OpenAtom Foundation)孵化及运营的开源项目,是一款由全球开发者共建的开源分布式操作系统。其目标是面向全场景、全连接、全智能时代,基于开源的方式搭建一个智能终端设备操作系统的框架和平台,促进万物互联产业的繁荣发展。从推出之日至今,OpenHarmony 操作系统的发展愈加迅速,生态系统建设愈加成熟,已经成为全球智能终端操作系统领域不可忽视的新生开源力量。

为了让更多的人了解并熟练使用 OpenHarmony 轻量系统,笔者将自身学习经历以案例的形式进行了梳理、总结,形成了本书,供更多的爱好者参考、学习。

本书特点

本书从基础知识着手,通过大量的案例全方面系统地讲解 OpenHarmony 轻量系统开发,由表及里、深入浅出、循序渐进,集前瞻性、应用性、趣味性于一体。主要面向的读者是对 OpenHarmony 轻量系统应用开发感兴趣的学生、开发人员或者相关从业人员,让更多的人掌握 OpenHarmony 轻量系统应用开发技术。书中的案例基于海思 Hi3861 的开发板编写。

第 1 章 OpenHarmony 轻量系统开发基础,主要讲解 OpenHarmony 发展历程、环境搭建、应用模块工程创建与运行、应用模块启动流程解析等内容,为后续开发做准备。

第 2 章 OpenHarmony 轻量系统设备开发,通过 25 个案例详细讲解 OpenHarmony 轻量系统核心设备接口 WatchDog、ADC、GPIO、PWM、I^{2}C 等开发技术。

第 3 章 OpenHarmony 轻量系统系统开发,通过 6 个案例详细讲解多任务、互斥锁、软件定时器、按键中断处理、内存申请与释放等开发技术。

第 4 章 OpenHarmony 轻量系统智能设备开发,通过 5 个案例详细讲解 OpenHarmony 轻量系统多设备、多模块等开发技术。

第 5 章 OpenHarmony 轻量系统物联网开发,通过 13 个案例详细讲解 OpenHarmony 轻量系统在物联网应用中的开发技术,涉及的技术有 WiFi 组网、cJSON、网络通信和物联网。

第 6 章综合案例:遥控小车,基于 OpenHarmony 和 HarmonyOS 实现,案例分为 HarmonyOS 手表端、服务器端和 OpenHarmony 开发板端这 3 端。从技术架构、功能实现、网络通信等方面进行讲解,提升读者 OpenHarmony 轻量系统应用开发技术综合开发能力。

扫描目录上方的二维码可下载本书源代码。

致谢

　　本书虽然倾注了笔者的全部努力,但由于水平有限,书中难免有疏漏之处,敬请广大读者谅解。感谢您购买本书,祝您读书快乐!

　　感谢出版社的所有工作人员在本书编写和审核过程中提供的无私帮助和宝贵建议,正是由于他们的耐心和支持才让本书得以出版。

戈 帅

2023 年 5 月

目录
CONTENTS

本书源代码

第1章 OpenHarmony 轻量系统开发基础 ………… 1

1.1 初识 OpenHarmony 操作系统 ………… 1
1.1.1 OpenHarmony 概述 ………… 1
1.1.2 OpenHarmony 操作系统特性 ………… 2

1.2 OpenHarmony 操作系统开发环境搭建 ………… 4
1.2.1 Linux 虚拟计算机环境搭建 ………… 4
1.2.2 Ubuntu 服务器环境搭建 ………… 11
1.2.3 OpenHarmony 编译环境搭建 ………… 31
1.2.4 Windows 开发环境搭建 ………… 48

1.3 OpenHarmony 轻量系统应用模块开发 ………… 61
1.3.1 应用模块的源码结构 ………… 61
1.3.2 模块初始化接口 ………… 62
1.3.3 应用模块开发 ………… 63
1.3.4 应用模块测试 ………… 65

1.4 OpenHarmony 轻量系统应用模块启动流程解析 ………… 68
1.4.1 应用模块启动流程解析 ………… 68
1.4.2 应用模块启动流程验证 ………… 71

第2章 OpenHarmony 轻量系统设备开发 ………… 74

2.1 案例1：WatchDog ………… 74
2.2 ADC ………… 78
2.2.1 案例2：雨滴探测器 ………… 80

2.2.2　案例 3：游戏杆 ··· 83
　　2.2.3　案例 4：烟雾探测器 ·· 85
　　2.2.4　案例 5：声音监测仪 ·· 88
　　2.2.5　案例 6：光照检测仪 ·· 90
　　2.2.6　案例 7：生命探测仪 ·· 92
　　2.2.7　案例 8：土壤湿度监测仪 ···································· 95
　　2.2.8　案例 9：电压调节器 ·· 97
2.3　GPIO ··· 99
　　2.3.1　案例 10：工作指示灯 ······································ 99
　　2.3.2　案例 11：智能开关 ······································· 102
　　2.3.3　案例 12：SOS 摩斯密码发射器 ······························ 105
　　2.3.4　案例 13：倾斜检测仪 ····································· 107
　　2.3.5　案例 14：地震监测仪 ····································· 109
　　2.3.6　案例 15：机械手臂 ······································· 111
　　2.3.7　案例 16：缝隙探测器 ····································· 113
　　2.3.8　案例 17：触摸感应器 ····································· 115
　　2.3.9　案例 18：火焰探测器 ····································· 117
　　2.3.10　案例 19：测距仪 ·· 119
2.4　PWM ·· 122
　　2.4.1　案例 20：自动门 ··· 124
　　2.4.2　案例 21：炫彩灯 ··· 127
　　2.4.3　案例 22：救护车警报器 ··································· 130
　　2.4.4　案例 23：音乐盒 ··· 132
2.5　I^{2}C ··· 135
　　2.5.1　案例 24：温湿度监测仪 ··································· 137
　　2.5.2　案例 25：电子阅读器 ····································· 143

第 3 章　OpenHarmony 轻量系统系统开发 ································ 147

3.1　任务 ··· 147
　　3.1.1　案例 26：计时器 ··· 147
　　3.1.2　案例 27：自动售票系统 V1.0 ······························· 149
3.2　案例 28：自动售票系统 V2.0 ······································ 151
3.3　案例 29：软件定时器 ·· 154
3.4　案例 30：按键中断处理 ·· 158
3.5　案例 31：内存申请与释放 ·· 162

第 4 章　OpenHarmony 轻量系统智能设备开发 ………………………… 166

4.1　案例 32：智能雨刷 ……………………………………………………… 166
4.2　案例 33：智能雷达 ……………………………………………………… 169
4.3　案例 34：智能人体感应灯 ……………………………………………… 173
4.4　案例 35：智能红外报警器 ……………………………………………… 175
4.5　案例 36：智能火焰报警器 ……………………………………………… 178

第 5 章　OpenHarmony 轻量系统物联网开发 ………………………… 182

5.1　WiFi 技术 ………………………………………………………………… 182
5.1.1　案例 37：STA 端点接入 ……………………………………… 182
5.1.2　案例 38：AP 站点创建 ………………………………………… 186
5.2　cJSON ……………………………………………………………………… 189
5.2.1　案例 39：cJSON 对象封装 …………………………………… 190
5.2.2　案例 40：cJSON 对象解析 …………………………………… 192
5.2.3　案例 41：cJSON 数组封装 …………………………………… 194
5.2.4　案例 42：cJSON 数组解析 …………………………………… 197
5.3　网络通信 ………………………………………………………………… 199
5.3.1　案例 43：UDP 客户端应用 …………………………………… 199
5.3.2　案例 44：UDP 服务器端应用 ………………………………… 204
5.3.3　案例 45：TCP 客户端应用 …………………………………… 208
5.3.4　案例 46：TCP 服务器端应用 ………………………………… 213
5.4　物联网 …………………………………………………………………… 218
5.4.1　案例 47：MQTT 第三方库移植 ……………………………… 218
5.4.2　案例 48：MQTT 协议应用 …………………………………… 223
5.4.3　案例 49：MQTT 物联网应用开发 …………………………… 229

第 6 章　综合案例：遥控小车 ……………………………………………… 235

6.1　案例介绍 ………………………………………………………………… 235
6.1.1　案例架构介绍 ………………………………………………… 235
6.1.2　技术架构图 …………………………………………………… 235
6.1.3　运行效果 ……………………………………………………… 236
6.2　OpenHarmony 开发板端功能实现 …………………………………… 237
6.2.1　MQTT 通信模块功能实现 …………………………………… 237
6.2.2　小车指令执行模块功能实现 ………………………………… 243
6.2.3　主模块功能实现及测试 ……………………………………… 248

6.3 手表端功能实现 …………………………………………………………… 251
　　6.3.1 创建并配置工程 ……………………………………………………… 251
　　6.3.2 UI设计与实现 ………………………………………………………… 255
　　6.3.3 功能实现 ……………………………………………………………… 262
6.4 多端联调 …………………………………………………………………… 269

第1章 OpenHarmony 轻量系统开发基础

从本章开始,我们就进入了 OpenHarmony 轻量系统开发的学习之旅。本章是全书的准备阶段,分为 4 节,分别为初识 OpenHarmony 操作系统、OpenHarmony 操作系统开发环境搭建、OpenHarmony 轻量系统应用模块开发和 OpenHarmony 轻量系统应用模块启动流程解析。

1.1 初识 OpenHarmony 操作系统

1.1.1 OpenHarmony 概述

OpenHarmony 操作系统(开源鸿蒙操作系统)是由开放原子开源基金会(OpenAtom Foundation)孵化及运营的开源项目,是一款由全球开发者共建的开源分布式操作系统。其目标是面向全场景、全连接、全智能时代、基于开源的方式,搭建一个智能终端设备操作系统的框架和平台,促进万物互联产业的繁荣发展。从推出之日至今,OpenHarmony 操作系统的发展愈加迅速,生态系统建设愈加成熟,已经成为全球智能终端操作系统领域不可忽视的新生开源力量。OpenHarmony 操作系统的基础功能由华为研发并捐献给开放原子开源基金会。

OpenHarmony 操作系统的主要发展历程如下:

(1) 2012 年 9 月,华为开始规划 OpenHarmony 操作系统。

(2) 2017 年 5 月,华为完成 OpenHarmony 操作系统内核 1.0 的技术验证。

(3) 2020 年 9 月,华为将 OpenHarmony v1.0 捐献给开放原子开源基金会,从此 OpenHarmony 操作系统由开放原子开源基金会孵化及运营。同月,开放原子开源基金会对 OpenHarmony v1.0 进行全量开源发布。

(4) 2021 年 6 月 1 日,开放原子开源基金会对 OpenHarmony v2.0 进行全量开源发布。

(5) 2021 年 9 月 30 日,开放原子开源基金会对 OpenHarmony v3.0 LTS(Long Time Support,长期支持)全量开源发布。

(6) 2021 年 10 月 27 日,Eclipse 基金会发布公告,宣布推出基于 OpenHarmony 的操

作系统 Oniro。

（7）2021 年 12 月 9 日，慧思睿通芯片研发团队成功将 OpenHarmony 3.0 系统移植到了龙芯 1C300 芯片上，成为 OpenHarmony 发展史上又一座里程碑。

1.1.2　OpenHarmony 操作系统特性

1. OpenHarmony 支持 3 种类型操作系统

（1）轻量系统（Mini System）。面向 MCU 类处理器，例如 ARM Cortex-M、RISC-V 32 位的设备，硬件资源极其有限，支持的设备的最小内存为 128KiB，可以提供多种轻量级网络协议，轻量级的图形框架，以及丰富的 IoT 总线读写部件等。可支撑的产品包括智能家居领域的连接类模组、传感器设备、穿戴类设备等。

（2）小型系统（Small System）。面向应用处理器，例如 ARM Cortex-A 的设备，支持的设备的最小内存为 1MiB，可以提供更高的安全能力、标准的图形框架、视频编解码的多媒体能力等。可支撑的产品包括智能家居领域的 IP Camera、电子猫眼、路由器及智慧出行领域的行车记录仪等。

（3）标准系统（Standard System）。面向应用处理器，例如 ARM Cortex-A 的设备，支持的设备的最小内存为 128MiB，可以提供增强的交互能力、3D GPU 及硬件合成能力、更多控件及效果更丰富的图形能力、完整的应用框架。可支撑的产品包括高端的冰箱显示屏、各种手机等。

2. OpenHarmony 技术架构

OpenHarmony 整体遵从分层设计，从下向上依次为内核层、系统服务层、框架层和应用层。系统功能按照"系统→子系统→功能/模块"逐级展开，在多设备部署场景下，支持根据实际需求裁剪某些非必要的子系统或功能/模块。OpenHarmony 技术架构如图 1-1 所示。

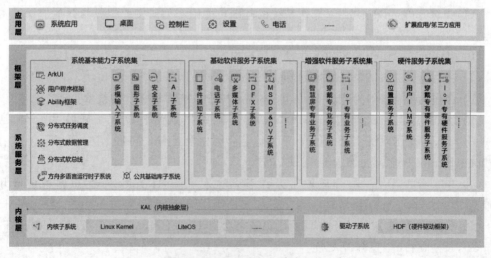

图 1-1　OpenHarmony 技术架构

1) 内核层

（1）内核子系统：采用多内核（Linux 内核或者 LiteOS）设计，支持针对不同资源受限设备选用适合的 OS 内核。内核抽象层（Kernel Abstract Layer，KAL）通过屏蔽多内核差异，对上层提供基础的内核能力，包括进程/线程管理、内存管理、文件系统、网络管理和外设管理等。

（2）驱动子系统：驱动框架（HDF）是系统硬件生态开放的基础，提供统一外设访问能力和驱动开发、管理框架。

2) 系统服务层

系统服务层是 OpenHarmony 的核心能力集合，通过框架层对应用程序提供服务。该层包含以下几部分。

（1）系统基本能力子系统集：为分布式应用在多设备上的运行、调度、迁移等操作提供了基础能力，由分布式软总线、分布式数据管理、分布式任务调度、公共基础库子系统、多模输入子系统、图形子系统、安全子系统、AI 子系统等组成。

（2）基础软件服务子系统集：提供公共的、通用的软件服务，由事件通知子系统、电话子系统、多媒体子系统、DFX（Design For X）子系统等组成。

（3）增强软件服务子系统集：提供针对不同设备的、差异化的能力增强型软件服务，由智慧屏专有业务子系统、穿戴专有业务子系统、IoT 专有业务子系统等组成。

（4）硬件服务子系统集：提供硬件服务，由位置服务子系统、用户 IAM 子系统、穿戴专有硬件服务子系统、IoT 专有硬件服务子系统等组成。

根据不同设备形态的部署环境，基础软件服务子系统集、增强软件服务子系统集、硬件服务子系统集的内部可以按子系统粒度裁剪，每个子系统的内部又可以按功能粒度裁剪。

3) 框架层

框架层为应用开发提供了 C/C++/JS 等多语言的用户程序框架和 Ability 框架，适用于 JS 语言的 ArkUI 框架，以及各种软硬件服务对外开放的多语言框架 API。根据系统的组件化裁剪程度，设备支持的 API 也会有所不同。

4) 应用层

应用层包括系统应用和第三方非系统应用。应用由一个或多个 FA（Feature Ability）或 PA（Particle Ability）组成，其中，FA 有 UI 界面，提供与用户交互的能力，而 PA 无 UI 界面，提供后台运行任务的能力及统一的数据访问抽象。基于 FA/PA 开发的应用，能够实现特定的业务功能，支持跨设备调度与分发，为用户提供一致、高效的应用体验。

3. OpenHarmony 技术特性

1) 硬件互助，资源共享

主要通过下列模块达成。

（1）分布式软总线：分布式软总线是多设备终端的统一基座，为设备间的无缝互联提供了统一的分布式通信能力，能够快速发现并连接设备，高效地传输任务和数据。

（2）分布式数据管理：分布式数据管理是在分布式软总线之上的能力，实现了应用程序数据和用户数据的分布式管理。用户数据不再与单一物理设备绑定,业务逻辑与数据存储分离,应用跨设备运行时数据无缝衔接,为打造一致、流畅的用户体验创造了基础条件。

(3) 分布式任务调度:分布式任务调度基于分布式软总线、分布式数据管理、分布式 Profile 等技术特性,构建统一的分布式服务管理(发现、同步、注册、调用)机制,支持对跨设备的应用进行远程启动、远程调用、绑定/解绑及迁移等操作,能够根据不同设备的能力、位置、业务运行状态、资源使用情况并结合用户的习惯和意图,选择最合适的设备运行分布式任务。

(4) 设备虚拟化:分布式设备虚拟化平台可以实现不同设备的资源融合、设备管理、数据处理,将周边设备作为该设备能力的延伸,共同形成一个超级虚拟终端。

2) 一次开发,多端部署

OpenHarmony 提供了用户程序框架、Ability 框架及 UI 框架,能够保证开发的应用在多终端运行时保证一致性,从而实现一次开发、多端部署。

多终端软件平台 API 具备一致性,确保用户程序的运行兼容性。

(1) 支持在开发过程中预览终端的能力适配情况,如 CPU、内存、外设、软件资源等。

(2) 支持根据用户程序与软件平台的兼容性来调度用户呈现。

3) 统一 OS,弹性部署

OpenHarmony 通过组件化和组件弹性化等设计方法,做到硬件资源的可大可小,在多种终端设备间按需弹性部署,全面覆盖了 ARM、RISC-V、x86 等各种 CPU,从数百 KiB 到 GiB 级别的 RAM。

1.2 OpenHarmony 操作系统开发环境搭建

本节讲解基于 Hi3861 的 OpenHarmony 轻量系统开发环境搭建,开发环境包含 Linux 编译服务器环境和 Windows 开发环境。

1.2.1 Linux 虚拟计算机环境搭建

本节讲解虚拟机软件 VirtualBox 的下载和安装、Ubuntu 20.04 镜像文件的下载和虚拟计算机的创建和配置,具体步骤如下。

1. 安装虚拟机软件 VirtualBox

在官方网站下载最新版本的 VirtualBox,官网网址为 https://www.virtualbox.org/,如图 1-2 所示。

双击已下载的安装包 VirtualBox-6.1.32-149290-Win.exe 进行默认安装,如图 1-3 所示。

2. 下载 Ubuntu 20.04 镜像文件

登录官方网站并下载 Ubuntu 20.04 镜像文件,官方网址为 https://releases.ubuntu.com/20.04/,如图 1-4 所示。

第1章　OpenHarmony轻量系统开发基础

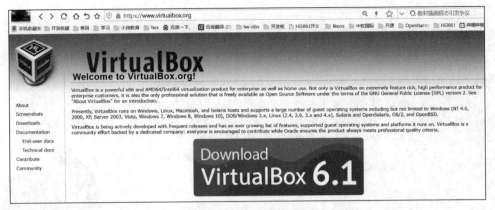

图 1-2　VirtualBox 官网

图 1-3　VirtualBox 安装包

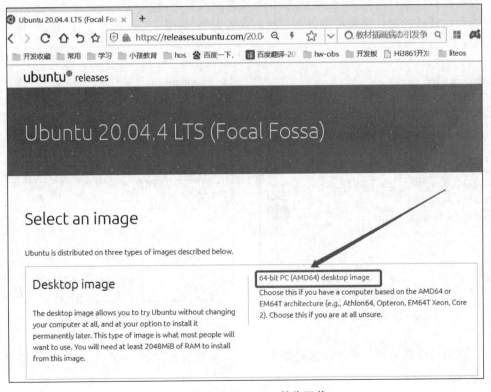

图 1-4　Ubuntu 20.04 镜像下载

3. 创建虚拟计算机

运行虚拟机软件 VirtualBox，单击"新建"按钮创建虚拟计算机，如图 1-5 所示。

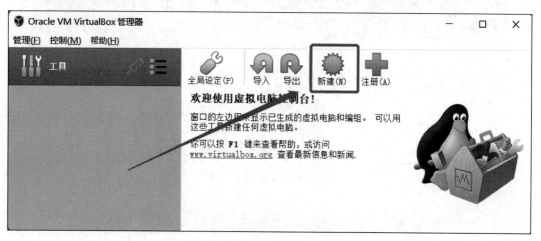

图 1-5　新建虚拟计算机

设置虚拟计算机名称、保存路径、操作系统类型及版本，如图 1-6 所示。

根据物理机的配置合理地设置虚拟机内存的大小，如图 1-7 所示。

图 1-6　设置名称和类型

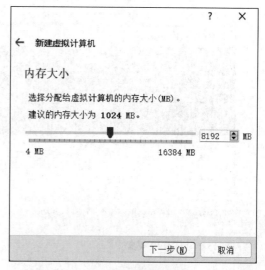

图 1-7　设置内存

选择虚拟硬盘创建方式并设置硬盘文件类型、分配方式、硬件大小及位置，默认即可，如图 1-8～图 1-11 所示。

4. 配置虚拟计算机

选择虚拟计算机，单击"设置"按钮，如图 1-12 所示。

第1章 OpenHarmony轻量系统开发基础 7

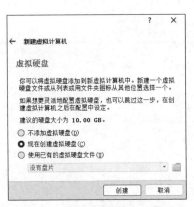

图 1-8 设置虚拟硬盘创建方式

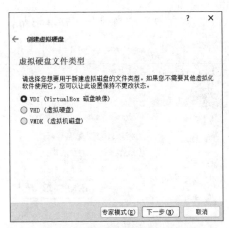

图 1-9 设置虚拟硬盘文件类型

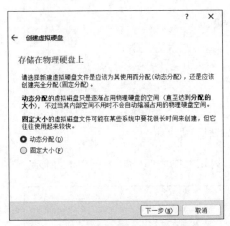

图 1-10 设置虚拟硬盘分配类型

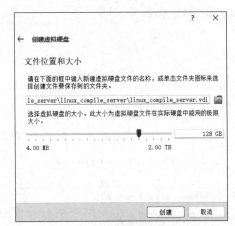

图 1-11 设置虚拟硬盘大小

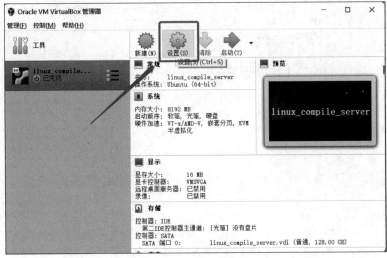

图 1-12 打开设置界面

依次选择"常规"→"高级"→"共享粘贴板"→"双向",启用"共享粘贴板"功能,如图 1-13 所示。

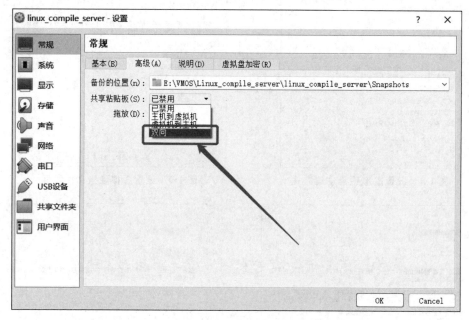

图 1-13　设置"共享粘贴板"功能

依次选择"常规"→"高级"→"拖放"→"双向",启用"拖放"功能,如图 1-14 所示。

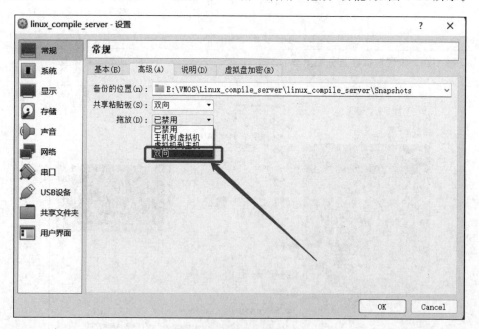

图 1-14　设置"拖放"功能

依次选择"系统"→"处理器"→"处理器数量",设置"处理器数量",如图 1-15 所示。

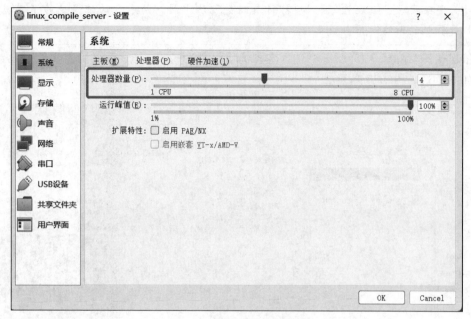

图 1-15　设置"处理器数量"

依次选择"网络"→"网卡 2"→"启用网络连接 "→"连接方式"→"仅主机(Host-Only)网络",启用"网卡 2",实现与 Windows 主机网络通信,如图 1-16 所示。

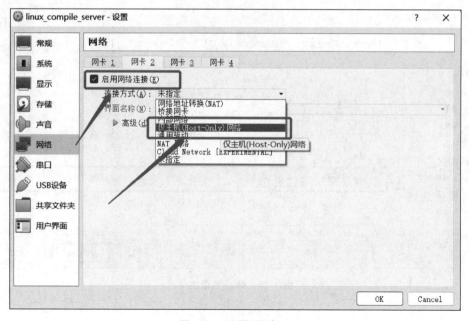

图 1-16　设置"网卡 2"

注意：不要在网卡1上设置，否则虚拟计算机会无法上网。网卡2的配置是为了实现主机与虚拟计算机的网络通信。

依次选择"存储"→"光驱"→"选择虚拟光盘文件"→"Ubuntu 系统镜像文件"，设置"光驱"，如图 1-17～图 1-19 所示。

图 1-17　设置光盘文件

图 1-18　选择光盘文件

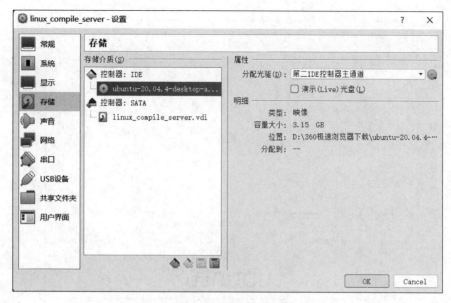

图 1-19 设置光驱镜像

1.2.2 Ubuntu 服务器环境搭建

本节讲解 Ubuntu 20.04 操作系统的安装和 Ubuntu 编译服务环境的配置,具体步骤如下。

1. 为虚拟计算机安装 Ubuntu 操作系统

单击"启动"按钮运行虚拟计算机,如图 1-20 和图 1-21 所示。

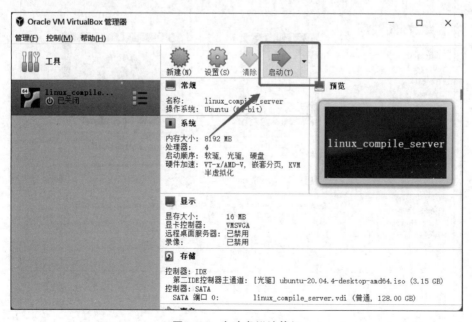

图 1-20 启动虚拟计算机

图 1-21 运行 Ubuntu 安装程序

单击 Install Ubuntu 按钮进行系统安装，如图 1-22 所示。

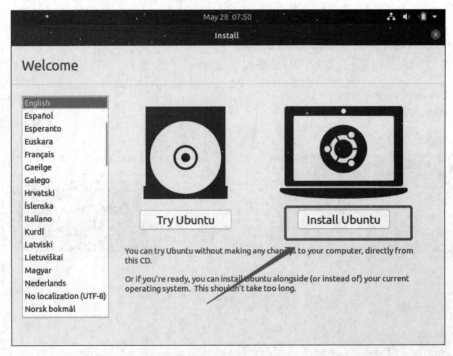

图 1-22 安装 Ubuntu

设置键盘布局,选择默认设置即可,如图 1-23 所示。

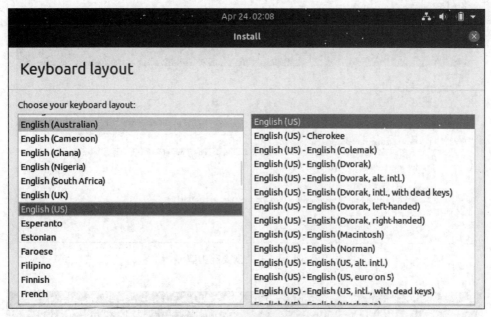

图 1-23　设置键盘布局

选择 Normal installation 进行正常软件安装,如图 1-24 所示。

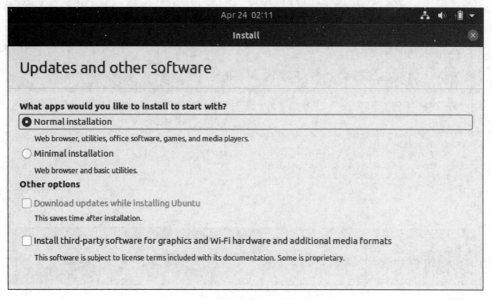

图 1-24　软件安装

选择 Erase disk and install Ubuntu 安装类型，清除磁盘内容并安装 Ubuntu，如图 1-25 和图 1-26 所示。

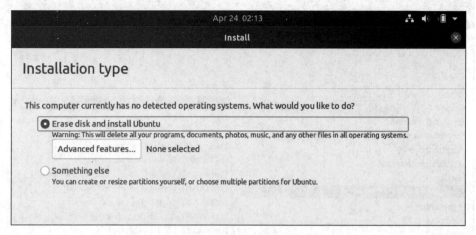

图 1-25　磁盘安装类型

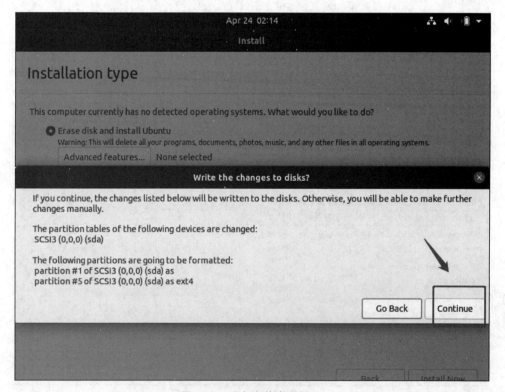

图 1-26　磁盘安装类型提示

设置用户名和虚拟计算机信息，根据个人情况输入相关用户名、计算机信息，如图1-27所示。

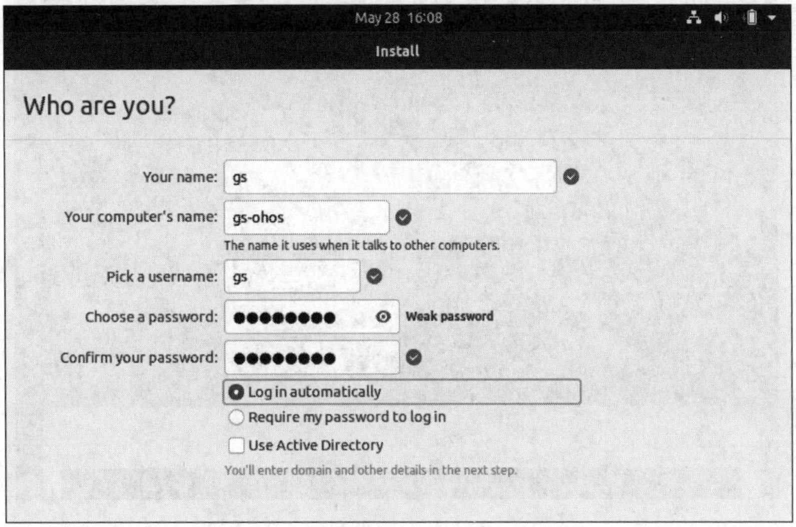

图 1-27　设置用户名和虚拟计算机信息

注意：一定要记住账号和密码，否则无法登录操作系统。

接下来会自动安装系统，同时在安装过程中会自动进行更新，所需时间比较长，系统安装完成后会提示重启，如图1-28～图1-30所示。

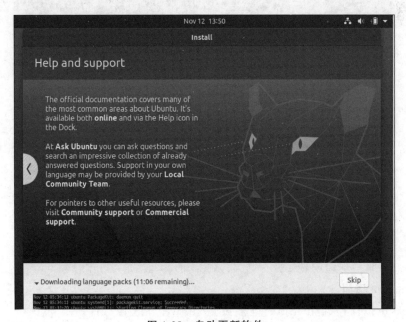

图 1-28　自动更新软件

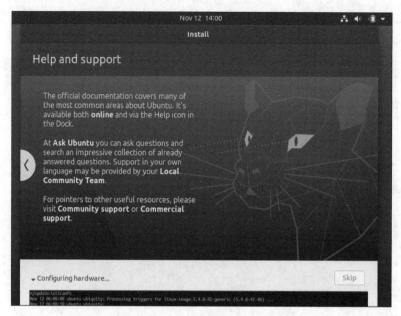

图 1-29 硬件配置

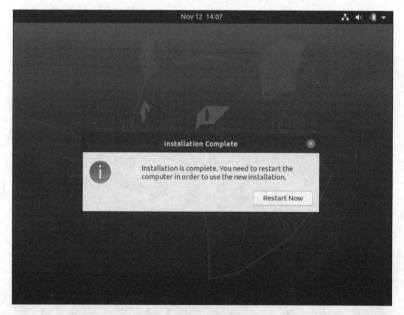

图 1-30 安装完成后提示重启

2．Ubuntu 服务环境配置

1）激活 root 账号

（1）重启虚拟计算机，按 Enter 键移除 Ubuntu 系统镜像文件。

（2）自动登录系统，如果系统安装时选择了 Require my password to log in，则需要输入用户名、密码登录系统，如图 1-31 所示。

图 1-31　系统用户登录界面

（3）第 1 次登录系统需要进行一些简单的配置，选择默认配置即可，如图 1-32 ～图 1-35 所示。

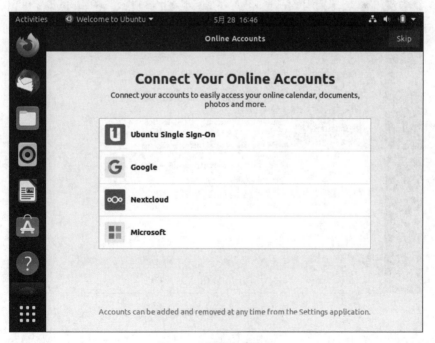

图 1-32　配置在线账号

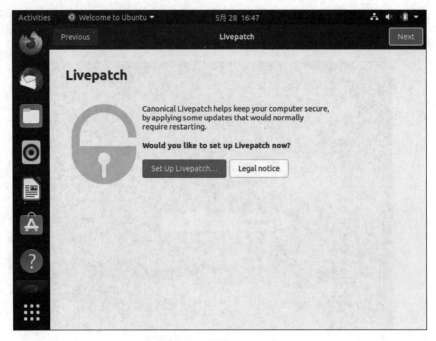

图 1-33　配置 Livepatch

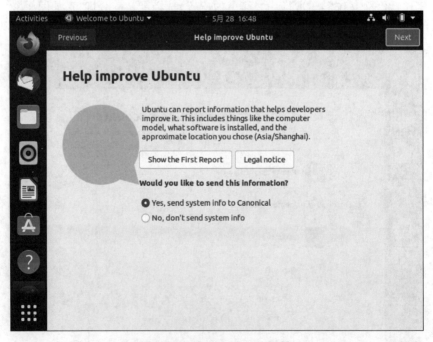

图 1-34　improve Ubuntu 配置

第1章 OpenHarmony轻量系统开发基础 | 19

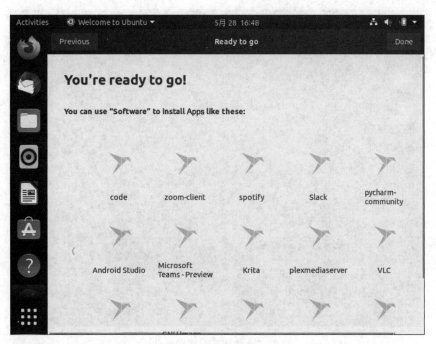

图 1-35 配置完成

(4) 在左侧快速启动栏中移除无用的 App，如图 1-36 和图 1-37 所示。

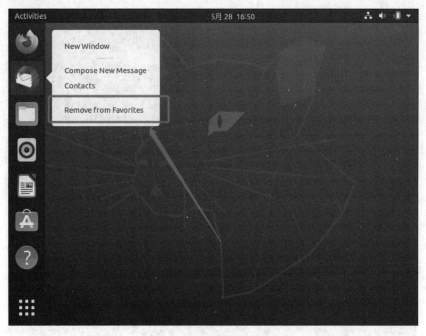

图 1-36 移除 Email 软件

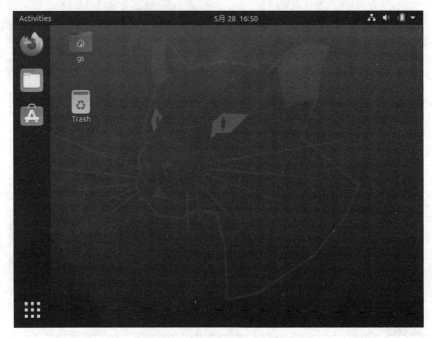

图 1-37 移除后的快速启动栏

（5）在左下角单击 ShowApplication，搜索 Terminal，然后单击运行 Terminal，如图 1-38 和图 1-39 所示。

图 1-38 搜索 Terminal

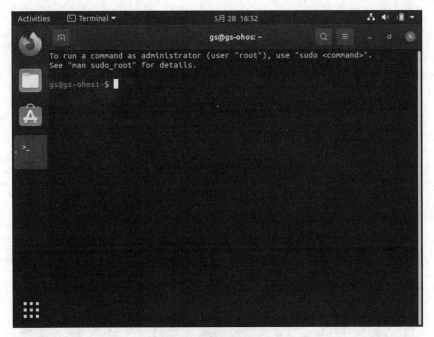

图 1-39 运行 Terminal

（6）在左侧快速启动栏右击终端软件 Terminal，然后添加到快速启动栏，如图 1-40 所示。

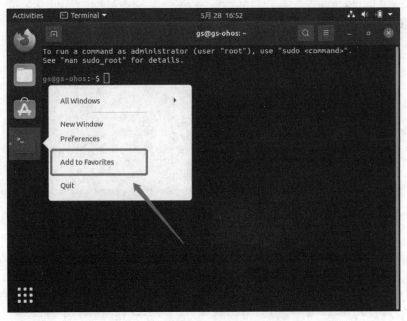

图 1-40 将 Terminal 添加到快速启动栏

（7）通过命令 sudo passwd 激活 root 账号，如图 1-41 所示。

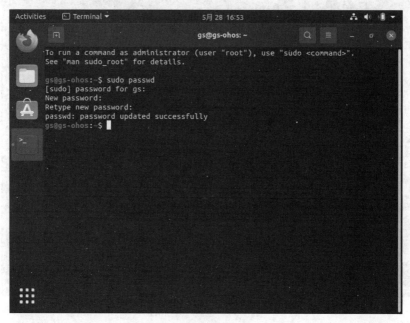

图 1-41　激活 root 账号

2）设置 Ubuntu 软件源服务器并更新

在 Show Application 中搜索并运行 Software&Updates，如图 1-42 所示。

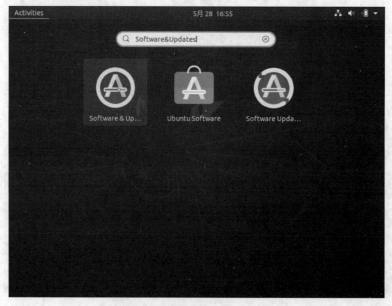

图 1-42　搜索 Software&Updates

3）选择中国的网速最快的软件源服务器

（1）单击 Download from 后面的下拉菜单，如图 1-43 所示。

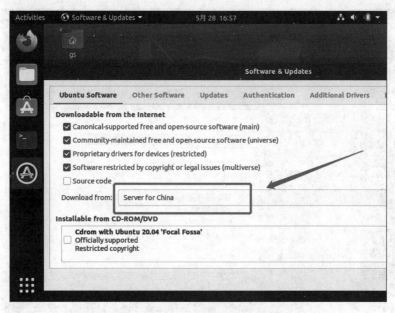

图 1-43　Software & Updates 设置

（2）选择 Other，进入软件源服务器选择界面，如图 1-44 所示。

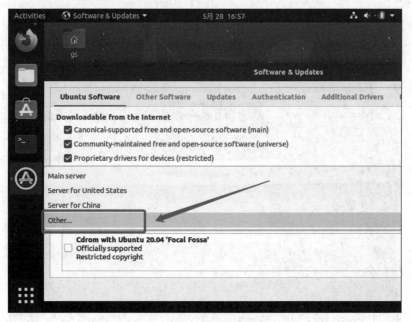

图 1-44　其他服务器配置

(3)准备测试中国的所有服务器,选择 China,然后单击 Select Best Server 按钮,测试得到网速最快的服务器,如图 1-45 所示。

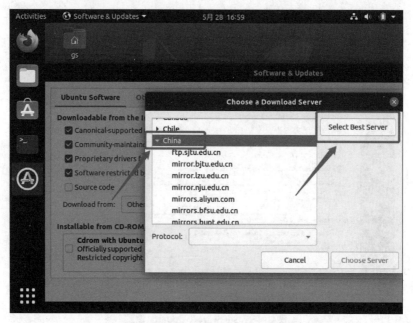

图 1-45　准备测试中国的所有服务器

(4)等待测试中国的网速最快的软件源服务器,如图 1-46 所示。

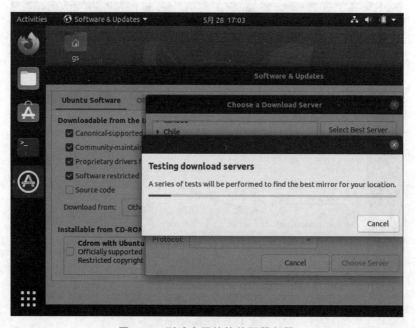

图 1-46　测试中国的软件源服务器

(5) 单击 Choose Server 选择测试得到的网速最快的服务器，如图 1-47 所示。

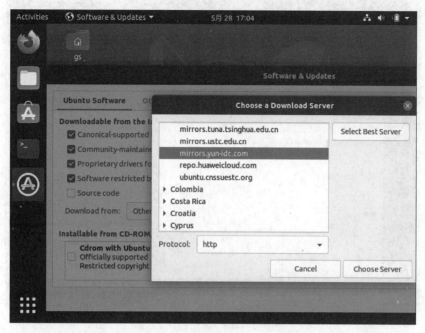

图 1-47　选择网速最快的服务器

(6) 输入超级管理员密码授权，如图 1-48 ～图 1-50 所示。

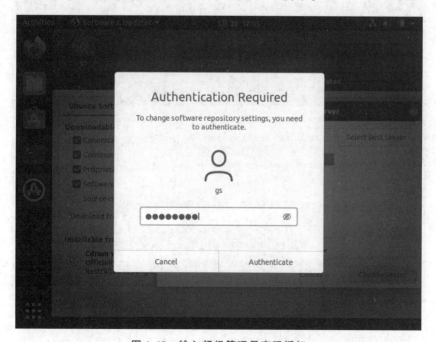

图 1-48　输入超级管理员密码授权

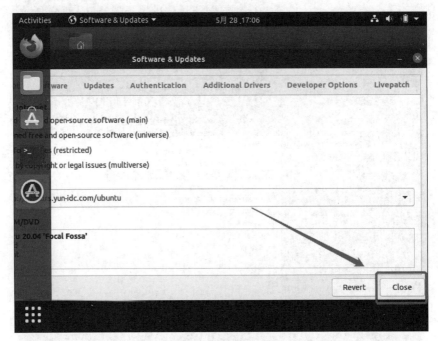

图 1-49　退出程序

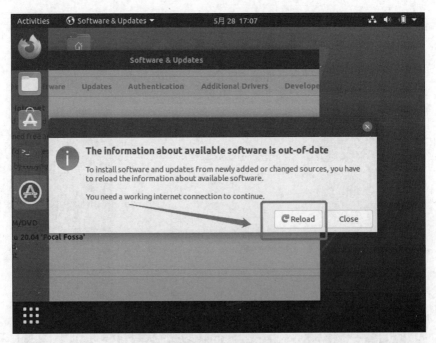

图 1-50　更新源

（7）在 Terminal 终端中输入命令 sudo apt-get update 进行软件源更新，如图 1-51 所示。

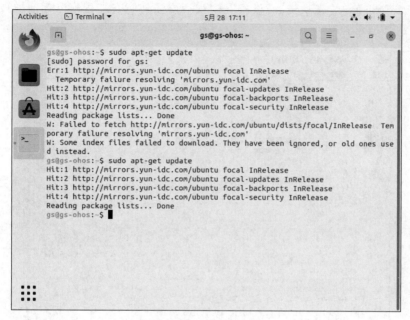

图 1-51　更新软件源

4）安装 net-tools 工具包

使用命令 sudo apt install net-tools 安装 net-tools 工具包（net-tools 工具包中包含了可以查看服务器 IP 地址的工具 ifconfig），如图 1-52 所示。

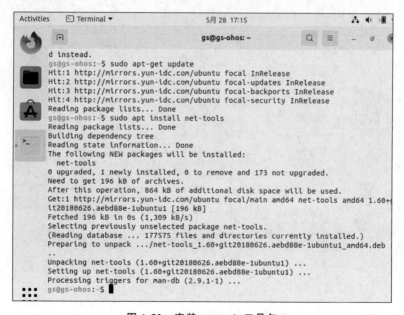

图 1-52　安装 net-tools 工具包

5）激活增强功能安装程序

（1）在虚拟机菜单中依次选择"设备"→"激活增强功能安装程序"，激活增强功能安装程序，如图1-53所示。

图 1-53　激活增强功能安装程序

（2）单击Run按钮，运行增强功能安装程序，如图1-54所示。

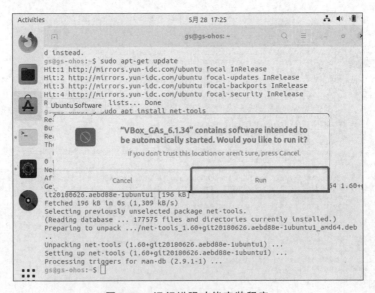

图 1-54　运行增强功能安装程序

（3）输入管理员密码授权完成增强功能的安装，如图 1-55 和图 1-56 所示。安装完成后启动虚拟计算机，增强功能生效后主机与虚拟计算机之间便可以进行双向复制、拖放。

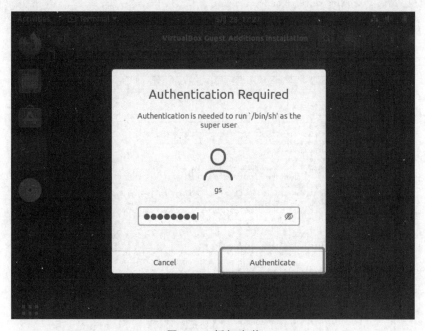

图 1-55　授权安装

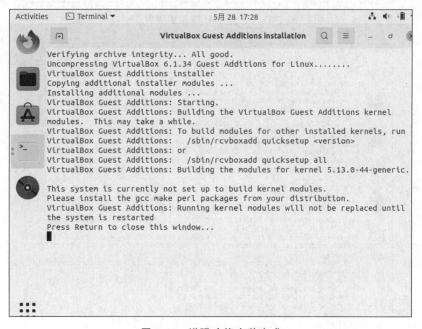

图 1-56　增强功能安装完成

(4) 安装并启动 SSH 服务，命令如下：

```
sudo apt-get install openssh-server -y
sudo service ssh start
```

运行效果如图 1-57 和图 1-58 所示。

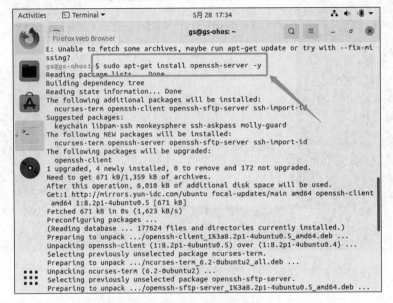

图 1-57　安装 SSH 服务器程序（一）

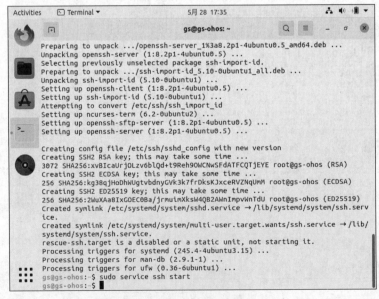

图 1-58　安装 SSH 服务器程序（二）

6) 安装 Samba 服务器程序

安装 Samba 服务器程序,命令如下:

```
sudo apt-get install samba -y
```

运行效果如图 1-59 所示。

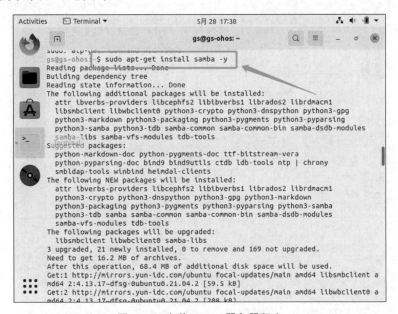

图 1-59　安装 Samba 服务器程序

1.2.3　OpenHarmony 编译环境搭建

本节讲解 OpenHarmony 系统源码下载和共享、OpenHarmony 轻量系统编译环境搭建和测试,具体步骤如下。

1. 共享鸿蒙操作系统源码

(1) 在用户根目录下创建文件夹 sharefolder,命令如下:

```
mkdir ~/sharefolder
```

(2) 进入 sharefolder 目录,下载 OpenHarmony code-1.0 源码,命令如下:

```
cd ~/sharefolder
wget https://repo.huaweicloud.com/harmonyos/os/1.0/code-1.0.tar.gz
```

运行效果如图 1-60 所示。

(3) 在 sharefolder 目录下创建 code-1.0 目录,将源码解压到 sharefolder/code-1.0,命令如下:

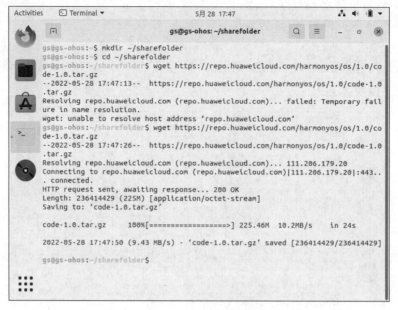

图 1-60　下载鸿蒙源码

```
mkdir code-1.0
tar -zxvf code-1.0.tar.gz -C code-1.0
```

运行效果如图 1-61 所示。

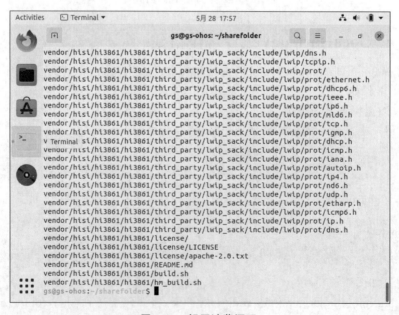

图 1-61　解压鸿蒙源码

（4）网络共享鸿蒙操作系统源码。右击 sharefolder 文件夹，选择 Local Network Share，进入设置网络共享文件夹界面，如图 1-62 所示。

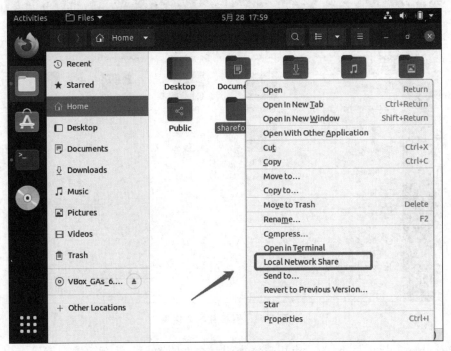

图 1-62　设置网络共享文件夹

设置 sharefolder 文件夹的访问权限，允许其他用户创建、删除文件，允许来宾用户访问文件夹。如果成功，则 sharefolder 文件夹上将会有一个绿色图标，如图 1-63 和图 1-64 所示。

图 1-63　设置网络共享文件夹

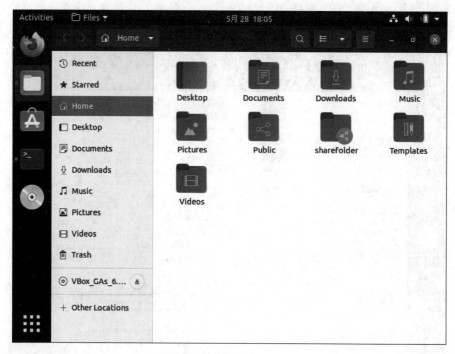

图 1-64 网络共享文件夹

进入 sharefolder 目录,为所有用户添加所有文件的读、写、执行权限。使所有用户都可以远程操作 sharefolder 目录下的文件和文件夹,命令如下:

```
cd ~/sharefolder
chmod -R 0777 *
```

在文件浏览器中输入共享文件夹地址,格式为虚拟机 IP 地址＋共享文件夹名,例如 \\192.168.56.105\sharefolder。如果可以正常打开并看到内容,则表示成功,如图 1-65 和图 1-66 所示。

在浏览器中输入共享文件夹的地址进行测试,例如 file://192.168.56.105/sharefolder/,如图 1-67 所示。

2. 安装交叉编译工具链及依赖

1) 安装 OpenHarmony 系统编译工具链及依赖

(1) 将 Linux Shell 修改为 dash,命令如下:

```
sudo dpkg-reconfigure dash
```

然后选择 No,如图 1-68 和图 1-69 所示。

第1章 OpenHarmony轻量系统开发基础

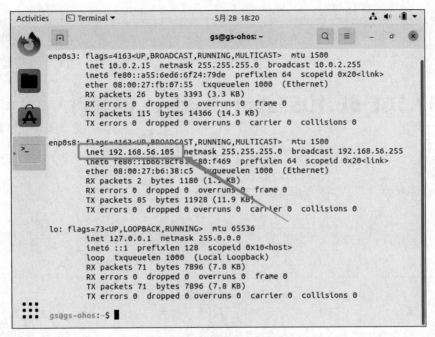

图 1-65 查看服务器 IP 地址

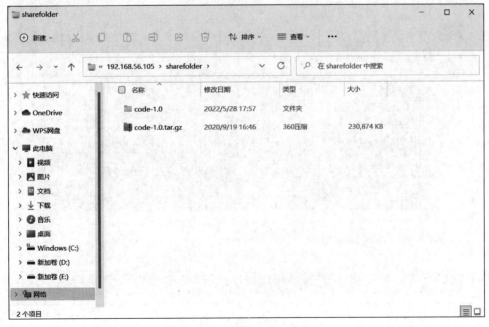

图 1-66 网络共享文件访问

图 1-67　浏览器访问网络共享文件夹

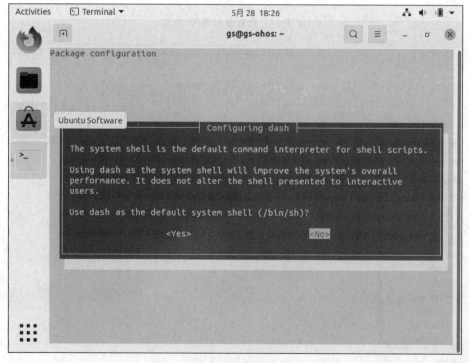

图 1-68　dash 配置

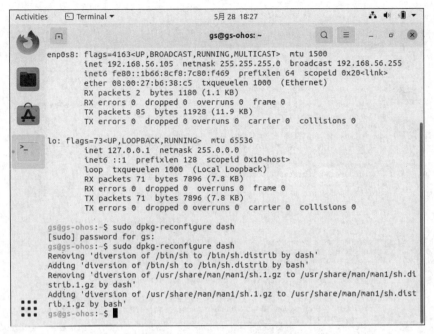

图 1-69　将 Linux Shell 修改为 dash

（2）安装 OpenHarmony 编译工具链的依赖工具和库（gcc、g++、make、zlib、libffi），命令如下：

sudo apt‐get install gcc ‐y&& sudo apt‐get install g++ ‐y&& sudo apt‐get install make ‐y&& sudo apt‐get install zlib* ‐y&& sudo apt‐get install libffi‐dev ‐y

运行效果如图 1-70 所示。

（3）安装 Python 3.8，命令如下：

sudo apt‐get install python3.8 ‐y

运行效果如图 1-71 所示。

（4）在/user/bin 下创建 Python 3.8 的软连接 Python，命令如下：

cd /usr/bin && sudo ln ‐s /usr/bin/python3.8 python && python ‐‐version

（5）安装 Python 包管理工具，命令如下：

sudo apt‐get install python3‐setuptools python3‐pip ‐y

运行效果如图 1-72 所示。

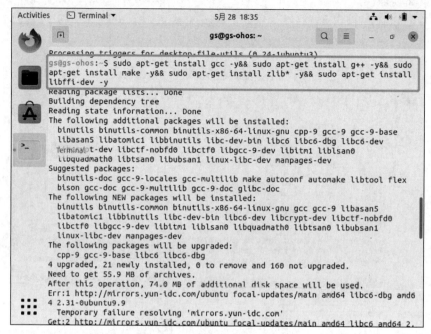

图 1-70 安装环境依赖工具和库

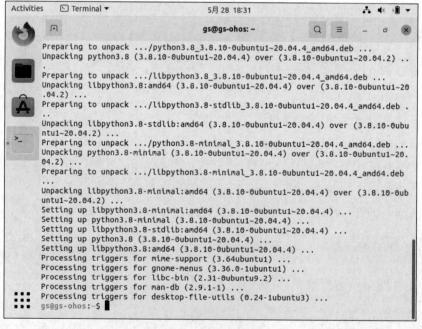

图 1-71 安装 Python 3.8

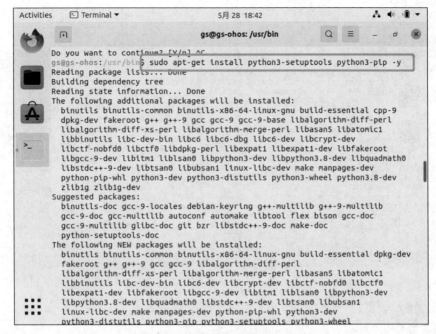

图 1-72　安装 Python 包管理工具

（6）安装 pip 工具，命令如下：

```
sudo pip3 install -- upgrade pip
```

运行效果如图 1-73 所示。

（7）安装 Python 模块 setuptools，命令如下：

```
sudo pip3 install setuptools
```

（8）安装 GUI menuconfig 工具（kconfiglib），命令如下：

```
sudo pip3 install kconfiglib
```

运行效果如图 1-74 所示。

（9）安装 pycryptodome，命令如下：

```
sudo pip3 install pycryptodome
```

运行效果如图 1-75 所示。

（10）安装 six 工具，命令如下：

```
sudo pip3 install six -- upgrade -- ignore- installed six
```

图 1-73 安装 pip 工具

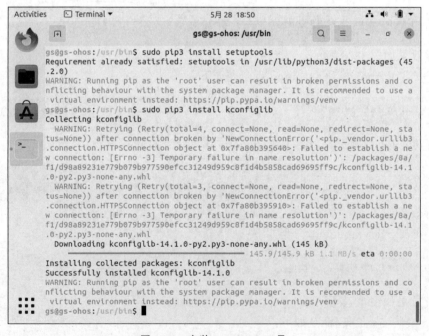

图 1-74 安装 kconfiglib 工具

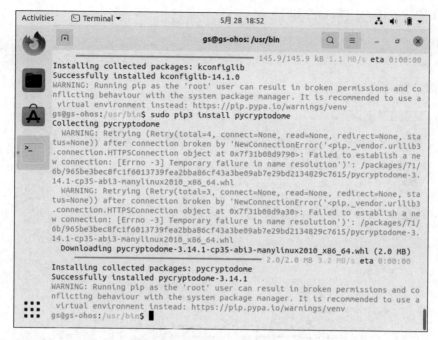

图 1-75　安装 pycryptodome 工具

运行效果如图 1-76 所示。

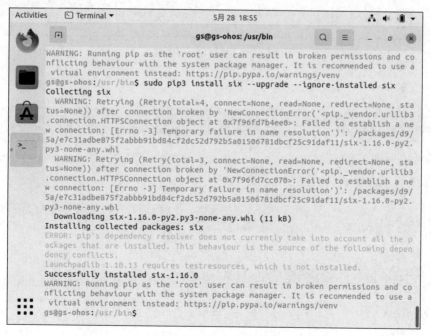

图 1-76　安装 six 工具

(11) 安装 ecdsa 工具，命令如下：

```
sudo pip3 install ecdsa
```

运行效果如图 1-77 所示。

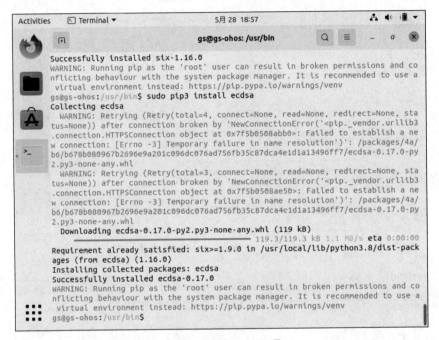

图 1-77　安装 ecdsa 工具

(12) 安装 scons 工具，命令如下：

```
sudo apt-get install scons
```

运行效果如图 1-78 所示。

(13) 安装 vim 工具，命令如下：

```
sudo apt install vim -y
```

运行效果如图 1-79 所示。

(14) 在当前用户目录下创建 tools 目录，然后进入 tools 目录，命令如下：

```
mkdir ~/tools
cd ~/tools
```

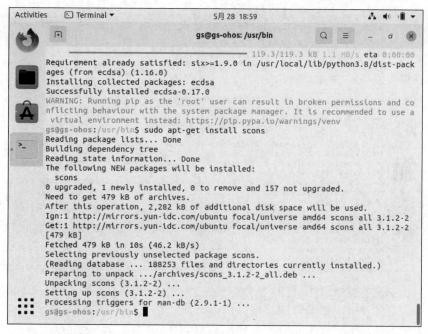

图 1-78　安装 scons 工具

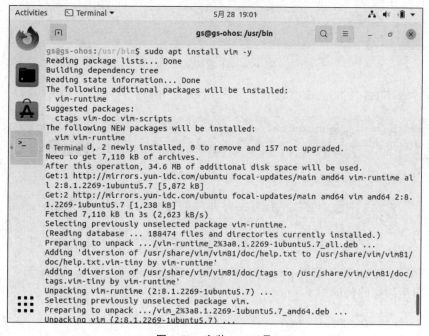

图 1-79　安装 vim 工具

(15) 下载并将 gn 解压到 tools 目录，命令如下：

```
wget https://repo.huaweicloud.com/harmonyos/compiler/gn/1523/linux/gn.1523.tar
tar -xvf gn.1523.tar -C ~/tools
```

运行效果如图 1-80 所示。

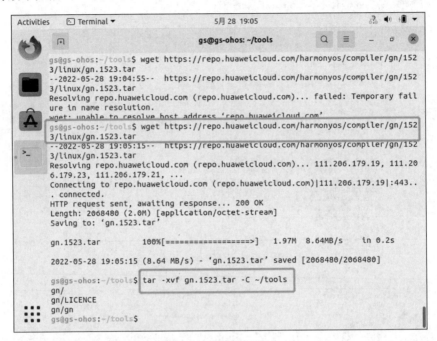

图 1-80　下载并解压 gn

(16) 下载并解压 ninja，命令如下：

```
wget https://repo.huaweicloud.com/harmonyos/compiler/ninja/1.9.0/linux/ninja.1.9.0.tar
tar -xvf ninja.1.9.0.tar -C ~/tools
```

(17) 下载并解压交叉编译工具 gcc_risc32，命令如下：

```
wget https://repo.huaweicloud.com/harmonyos/compiler/gcc_riscv32/7.3.0/linux/gcc_riscv32-linux-7.3.0.tar.gz
tar -xvf gcc_riscv32-linux-7.3.0.tar.gz -C ~/tools
```

(18) 在 ~/.bashrc 中配置 gn、ninja、gcc_riscv32 环境变量，使用 vim 编辑 ~/.bashrc，在文件的末尾添加的代码如下：

```
export PATH=~/tools/gn:$PATH
export PATH=~/tools/ninja:$PATH
export PATH=~/tools/gcc_riscv32/bin:$PATH
```

添加效果如图 1-81 所示。

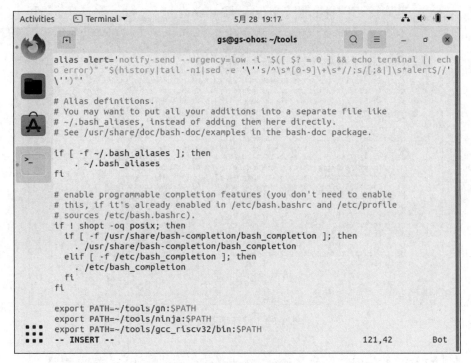

图 1-81　环境变量配置

（19）使环境变量生效，命令如下：

```
source ~/.bashrc
```

运行效果如图 1-82 所示。

3．编译环境测试

（1）进入 OpenHarmony 源码根目录，命令如下：

```
cd ~/sharefolder/code-1.0
```

运行效果如图 1-83 所示。

（2）编译系统源码，命令如下：

```
python build.py wifiiot
```

编译成功后会出现 BUILD SUCCESS，如图 1-84 和图 1-85 所示。

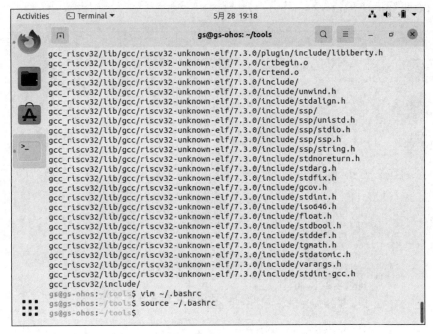

图 1-82　使环境变量生效

图 1-83　将当前路径更改为 code-1.0

图 1-84　运行命令编译系统源码

图 1-85　系统源码编译成功

1.2.4　Windows 开发环境搭建

本节讲解在 Windows 环境下开源鸿蒙操作系统源码映射与断开、编辑环境搭建和其他工具的安装及使用,具体内容如下。

1. 鸿蒙操作系统源码网络映射与断开

1) 创建网络驱动器前的准备工作

(1) 使用 ifconfig 命令查看当前系统的 IP 地址,如图 1-86 所示。

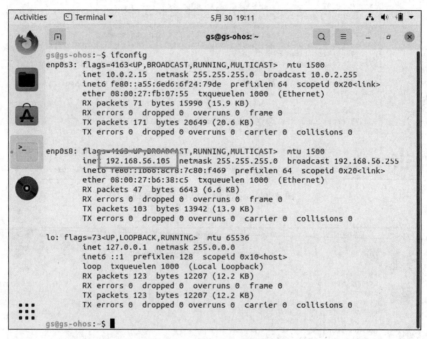

图 1-86　查询系统的 IP 地址

(2) 打开文件浏览器,查看共享文件夹名,如图 1-87 所示。

2) 创建网络驱动器

(1) 通过"此电脑"或其他文件夹打开文件浏览器,如图 1-88 所示。

(2) 右击"此电脑",选择"映射网络驱动器",如图 1-89 所示。

(3) 设置驱动器及网络共享文件夹路径,格式为虚拟机 IP 地址+共享文件夹名,例如 \\192.168.56.105\sharefolder,如图 1-90 所示。

(4) 单击"完成"按钮,完成创建操作,效果如图 1-91 所示。

3) 删除网络驱动器

在关闭 Ubuntu 编译服务器之前,先删除网络驱动器,否则无法正常使用文件浏览器。删除网络驱动器的步骤如下:

(1) 打开文件浏览器,如图 1-92 所示。

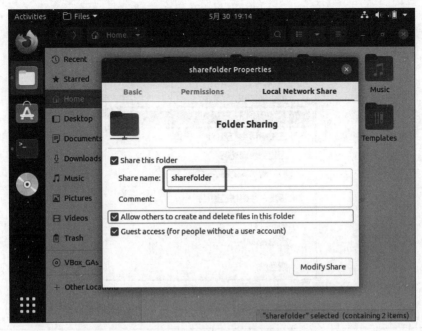

图 1-87　查看共享文件夹名

图 1-88　文件浏览器

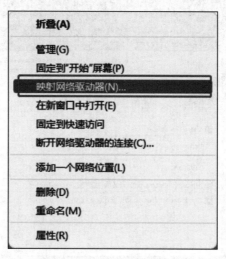

图 1-89 创建映射网络驱动器

图 1-90 映射网络驱动器

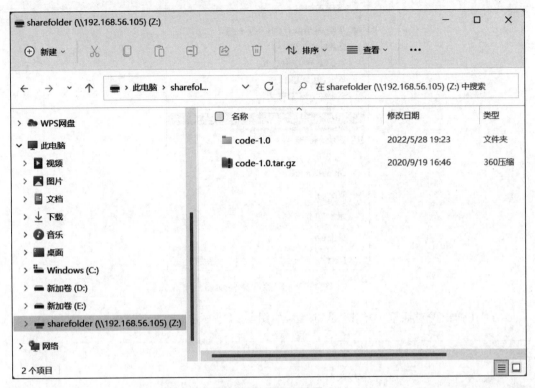

图 1-91 网络驱动器

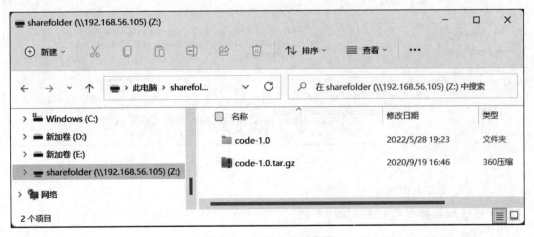

图 1-92 打开文件浏览器

（2）右击 sharefolder(\\192.168.56.105)(Z:)，然后选择"断开连接"，如图 1-93 所示。

图 1-93　删除网络驱动器

（3）在弹出的对话框中单击"是"按钮，如图 1-94 所示。

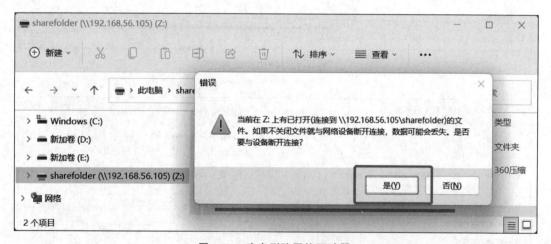

图 1-94　确定删除网络驱动器

（4）关闭所有文件浏览器，如图 1-95 所示。

2．搭建鸿蒙操作系统源码编辑环境

1）安装 Visual Studio Code

（1）登录 Visual Studio Code 官方网站下载最新版本的 Visual Studio Code，官方网址为 https://code.visualstudio.com/，如图 1-96 所示。

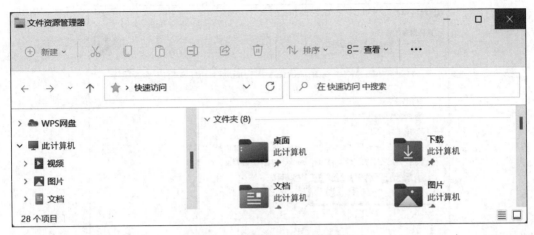

图 1-95　关闭所有文件浏览器

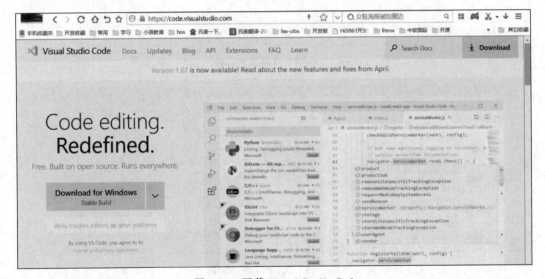

图 1-96　下载 Visual Studio Code

（2）双击运行安装包，进行默认安装，在安装过程中，当显示如下界面时，勾选"添加到 PATH（重启后生效）"和"创建桌面快捷方式"，如图 1-97 所示。

2）**安装鸿蒙操作系统源码编辑所需的插件**

（1）运行 Visual Studio Code，选择左侧工具栏中的 Extensions，在搜索栏中输入 C/C++，找到 C/C++ 插件并单击 Install 按钮进行安装，如图 1-98 所示。

（2）安装 GN 插件，如图 1-99 所示。

3）**Visual Studio Code 加载开源鸿蒙操作系统源码**

单击 Open Folder，选择开源鸿蒙操作系统源码目录，导入开源鸿蒙操作系统源码，如图 1-100 和图 1-101 所示。

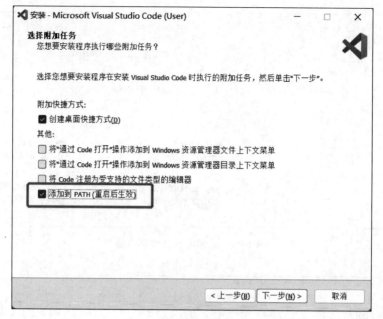

图 1-97 添加到 PATH

图 1-98 安装 C/C++ 插件

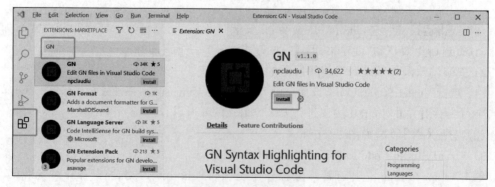

图 1-99 安装 GN 插件

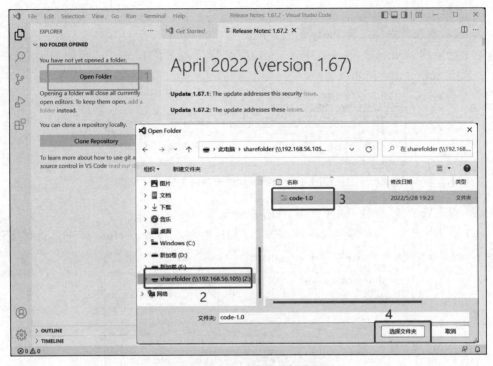

图 1-100　导入鸿蒙操作系统源码

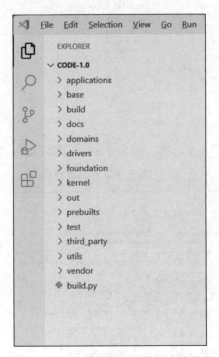

图 1-101　导入鸿蒙操作系统源码的效果

3. 其他工具安装测试

1) 远程登录客户端工具 Putty 的安装测试

(1) 登录网站 https://www.putty.org/，选择下载 64-bit x86 平台的安装包，下载 Putty 工具，如图 1-102 所示。

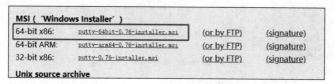

图 1-102　Putty 软件安装包下载

(2) 双击安装包，安装 Putty 工具，选择默认安装即可。

(3) 在 Host Name 处输入 Linux 服务器的 IP 地址，在 Connection type 下选择 SSH，单击 Open 按钮连接服务器，如图 1-103 和图 1-104 所示。

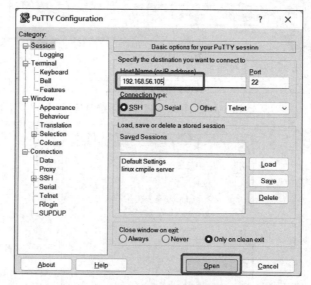

图 1-103　连接 SSH 服务器

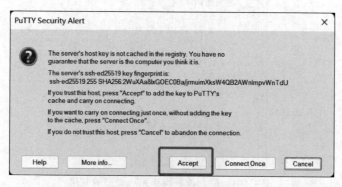

图 1-104　Putty 安全提示

（4）在 Putty 工具中输入账号、密码，登录 Ubuntu 服务器，如图 1-105 所示。

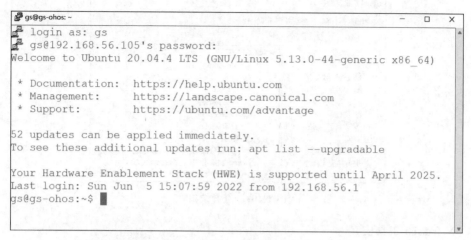

图 1-105　用户登录

2）安装开发板串口驱动

（1）连接 J3、J4 跳线使能串口，如图 1-106 所示。

图 1-106　连接 J3、J4 跳线

（2）使用数据线连接开发板与计算机。

（3）双击运行串口驱动 CH341SER，进行驱动安装，如图 1-107 和图 1-108 所示。

 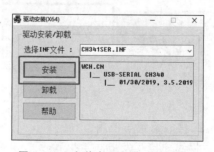

图 1-107　串口 CH341SER 驱动安装包　　图 1-108　安装串口驱动 CH341SER

（4）在设备管理器中可以看到 USB-SERIAL CH340 说明串口驱动安装成功，如图 1-109 所示。

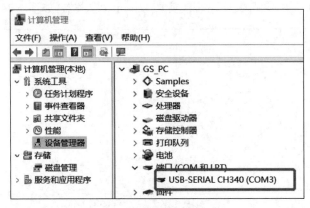

图 1-109　查看端口

3）NetAssist 工具安装

登录网站 https://soft.3dmgame.com/down/213757.html，下载 NetAssist 工具，如图 1-110 所示。

图 1-110　下载 NetAssist 工具

4）HiBurn 工具的配置与使用

（1）使用数据线连接开发板与主机。

（2）在"设备管理器"中查看开发板使用的端口号，如图 1-109 所示。

（3）运行 HiBurn，依次单击 Setting→COM settings，根据情况进行波特率设置，如图 1-111 所示。

图 1-111 波特率设置

（4）选择开发板使用的 COM 端口，如图 1-112 所示。

（5）勾选 Auto burn，如图 1-112 所示。

（6）选择烧写开源鸿蒙固件，固件的存储位置为\\out\wifiiot\Hi3861_wifiiot_app_allinone.bin，如图 1-112 所示。

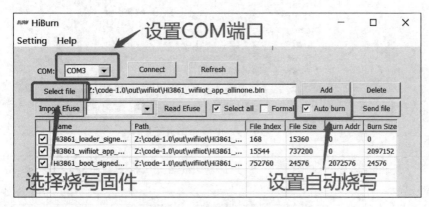

图 1-112 HiBurn 工具设置

（7）单击 Connect 按钮，如图 1-113 所示。

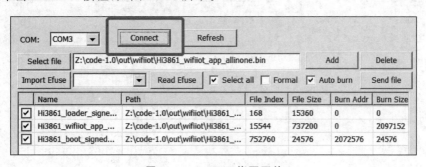

图 1-113 HiBurn 烧写固件

（8）单击开发板上的复位键 RST，进行固件烧写。

（9）当 3 个文件烧写完成时，单击 Disconnect 按钮断开连接。

5）IPOP 工具的配置与使用

（1）使用数据线连接开发板。

（2）在设备管理器中查看开发板使用的端口。

（3）依次单击"终端工具"→"新建连接"，选择"类型"对应的 COM 端口连接开发板，如图 1-114 所示。

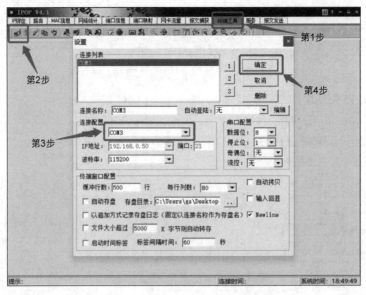

图 1-114　IPOP 终端工具配置

注意：如果开发板连接失败，则可能是 COM 端口被占用，单击 HiBurn 的 Disconnect 按钮释放 COM 端口即可。

（4）单击开发板上的 RST 复位键，运行效果如图 1-115 所示。

图 1-115　IPOP 终端工具连接成功

1.3 OpenHarmony 轻量系统应用模块开发

本节讲解 OpenHarmony 轻量系统的应用模块的源码结构、开发方法和测试方法,以及操作步骤。

1.3.1 应用模块的源码结构

OpenHarmony 系统功能可以按照"系统→子系统→功能/模块"逐级展开,系统是由一个或多个子系统组成的;子系统是由一个或多个模块构成的。模块可以独立构建,以二进制方式集成,是具备独立验证能力的二进制单元。一个模块一般实现一个特定的功能,模块源码包含一个模块构建脚本文件 BUILD.gn 和一个或多个功能源码文件。例如实现一个打印输出 Hello OpenHarmony introduction to master 50 cases 的功能的模块,模块源码只需包含模块构建脚本文件 BUILD.gn 和一个功能源码文件 hello_world.c,如图 1-116 所示,运行效果如图 1-117 所示。

图 1-116　应用模块结构

```
ready to OS start
sdk ver:Hi3861V100R001C00SPC025 2020-09-03 18:10:00
formatting spiffs...
FileSystem mount ok.
wifi init success!

Hello Openharmony introduction to master 50 cases

00 00:00:00 0 132 D 0/HIVIEW: hilog init success.
00 00:00:00 0 132 D 0/HIVIEW: log limit init success.
00 00:00:00 0 132 I 1/SAMGR: Bootstrap core services(count:3).
00 00:00:00 0 132 I 1/SAMGR: Init service:0x4ae53c TaskPool:0xfa1e4
00 00:00:00 0 132 I 1/SAMGR: Init service:0x4ae560 TaskPool:0xfa854
00 00:00:00 0 132 I 1/SAMGR: Init service:0x4ae6a8 TaskPool:0xfaa14
00 00:00:00 0 164 I 1/SAMGR: Init service 0x4ae560 <time: 0ms> succes
00 00:00:00 0 64 I 1/SAMGR: Init service 0x4ae53c <time: 0ms> success
00 00:00:00 0 8 D 0/HIVIEW: hiview init success.
00 00:00:00 0 8 I 1/SAMGR: Init service 0x4ae6a8 <time: 0ms> success!
00 00:00:00 0 8 I 1/SAMGR: Initialized all core system services!
00 00:00:00 0 64 I 1/SAMGR: Bootstrap system and application services
00 00:00:00 0 64 I 1/SAMGR: Initialized all system and application se
00 00:00:00 0 64 I 1/SAMGR: Bootstrap dynamic registered services(cou
```

图 1-117　模块运行效果

1.3.2 模块初始化接口

在系统运行中,模块是以库的形式加载到内存的,模块中的函数在没有其他模块调用之前不会主动执行,如果想执行模块中的指定函数,则需要用宏进行初始化函数,用来初始化函数的宏声明在头文件//utils\native\lite\include\ohos_init.h 中,不同的宏可以将函数注册到不同的阶段,系统在初始化该阶段时,函数会被调用执行,可以初始化的阶段如下。

1. CORE 阶段

系统在初始化 CORE 阶段时调用执行模块函数,用于初始化函数的宏有以下两个。

(1) CORE_INIT(func):使用默认优先级,默认优先级为 2。

(2) CORE_INIT_PRI(func,priority):可以通过 priority 参数设置优先级,优先级的取值为[0,4]。

2. SYS_SERVICE 阶段

系统在初始化 SYS_SERVICE 阶段时调用执行模块函数,用于初始化函数的宏有以下两个。

(1) SYS_SERVICE_INIT(func):使用默认优先级,默认优先级为 2。

(2) SYS_SERVICE_INIT_PRI(func,priority):可以通过 priority 参数设置优先级,优先级的取值为[0,4]。

3. SYS_FEATURE 阶段

系统在初始化 SYS_FEATURE 阶段时调用执行模块函数,用于初始化函数的宏有以下两个。

(1) SYS_FEATURE_INIT(func):使用默认优先级,默认优先级为 2。

(2) SYS_FEATURE_INIT_PRI(func,priority):可以通过 priority 参数设置优先级,优先级的取值为[0,4]。

4. RUN 阶段

系统在初始化 RUN 阶段时调用执行模块函数,用于初始化函数的宏有以下两个。

(1) SYS_RUN(func):使用默认优先级,默认优先级为 2。

(2) SYS_RUN_PRI(func,priority):可以通过 priority 参数设置优先级,优先级的取值为[0,4]。

5. APP_SERVICE 阶段

系统在初始化 APP_SERVICE 阶段时调用执行模块函数,用于初始化函数的宏有以下两个。

(1) SYSEX_SERVICE_INIT(func):使用默认优先级,默认优先级为 2。

(2) SYSEX_SERVICE_INIT_PRI(func,priority):可以通过 priority 参数设置优先级,优先级的取值为[0,4]。

6. APP_FEATURE 阶段

系统在 APP_FEATURE 阶段初始化时调用执行模块函数,用于初始化函数的宏有以下4个。

(1) SYSEX_FEATURE_INIT(func)：使用默认优先级，默认优先级为2。

(2) SYSEX_FEATURE_INIT_PRI(func, priority)：可以通过 priority 参数设置优先级，优先级的取值为[0,4]。

(3) APP_FEATURE_INIT(func)：使用默认优先级，默认优先级为2。

(4) APP_FEATURE_INIT_PRI(func, priority)：可以通过 priority 参数设置优先级，优先级的取值为[0,4]。

7. APP_SERVICE 阶段

系统在初始化 APP_SERVICE 阶段时调用执行模块函数，用于初始化函数的宏有以下两个。

(1) APP_SERVICE_INIT(func)：使用默认优先级，默认优先级为2。

(2) APP_SERVICE_INIT_PRI(func, priority)：可以通过 priority 参数设置优先级，优先级的取值为[0,4]。

1.3.3 应用模块开发

知道了应用模块的结构及初始接口，接下来以最经典的 Hello World 来讲解应用模块的开发。首先，创建工程目录及文件；其次，功能实现；再次，编写模块构建脚本；最后，将模块配置到应用子系统。

具体步骤如下：

1. 创建本书源码目录 ohos_50

在应用子系统目录//applications/sample/wifi-iot/app 中，创建目录 ohos_50，用来存放本书所有章节的源码，如图 1-118 所示。

图 1-118 本书源码目录

注意：如果创建文件夹失败，则有可能是文件夹缺少当前用户的操作权限，应在 code-1.0 目录下使用命令 chmod -R 0777 * 修改权限。

2. 创建应用模块工程目录 chapter_01

在 ohos_50 目录下创建应用模块 ch_01_hello_world 的工程目录 chapter_01，如图 1-119 所示。

3. 创建应用模块源码文件 hello_world.c

在应用模块 ch_01_hello_world 的工程目录 chapter_01 下，创建源码文件 hello_world.c，如图 1-120 所示。

4. 引入依赖的头文件

引用函数 printf 和宏 SYS_RUN 所依赖的头文件，代码如下：

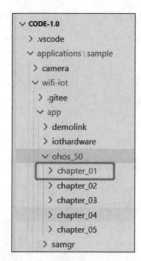 

图 1-119　应用模块工程目录　　图 1-120　源码文件 hello_world.c

```
#include <stdio.h>
#include <ohos_init.h>
```

5. 编写功能函数 HelloWorld

编写功能函数 HelloWorld，在函数中调用 printf 函数完成打印输出 Hello World 功能，代码如下：

```
static void HelloWorld(void)
{
    //函数 printf 声明在 stdio.h 头文件中,其功能是打印输出格式化字符串
    printf("\r\nHello World\n\r");
}
```

6. 初始化函数 HelloWorld

使用宏 SYS_RUN 初始化 HelloWorld 函数，使 HelloWorld 函数在系统初始化 RUN 阶段被执行，代码如下：

```
//宏 SYS_RUN 声明在 ohos_init.h 文件中,用来初始化模块入口函数
SYS_RUN(HelloWorld);
```

7. 创建并编写模块构建脚本 BUILD.gn

在模块目录 chapter_01 下，创建模块构建脚本文件 BUILD.gn，如图 1-121 所示。
编写模块构建脚本文件 BUILD.gn，指定模块名称(静态库名称)和依赖的源码文件，代码如下：

图 1-121　编译脚本文件 BUILD.gn

```
//applications/sample/wifi-iot/app/ohos_50/chapter_01/BUILD.gn
#将生成的模块名称指定为 ch_01_hello_world
static_library("ch_01_hello_world") {
    #模块依赖的源码文件,当有多个文件时使用逗号隔开
    sources = [
        "hello_world.c",
    ]
}
```

8. 将模块配置到应用子系统

到目前为止,如果编译系统源码,则模块 ch_01_hello_world 的源码不会参与编译,需要在应用子系统的编译构建脚本 BUILD.gn 中配置模块 ch_01_hello_world,具体的操作是在 features 中添加一条记录,格式为"模块目录的相对路径"+":"+"模块名称",代码如下:

```
//applications/sample/wifi-iot/app/BUILD.gn
import("//build/lite/config/component/lite_component.gni")

lite_component("app") {
    features = [
        "ohos_50/chapter_01:ch_01_hello_world",
    ]
}
```

到目前为止应用模块开发完成了,在 1.3.4 节将会讲解应用模块的编译、烧写、测试。

1.3.4　应用模块测试

本节讲解系统源码的编译、固件的烧写及测试,具体内容如下。

1. 编译系统源码

(1) 运行终端工具 Putty,连接 Linux 编译服务器,进行用户登录。

(2) 将当前目录切换为 code-1.0,编译系统源码,命令如下:

```
cd sharefolder/code-1.0/
python build.py wifiiot
```

运行效果如图 1-122 和图 1-123 所示。

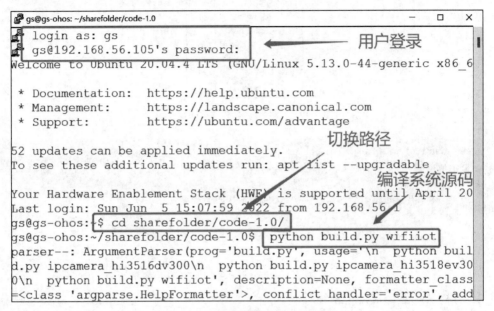

图 1-122 编译鸿蒙操作系统源码

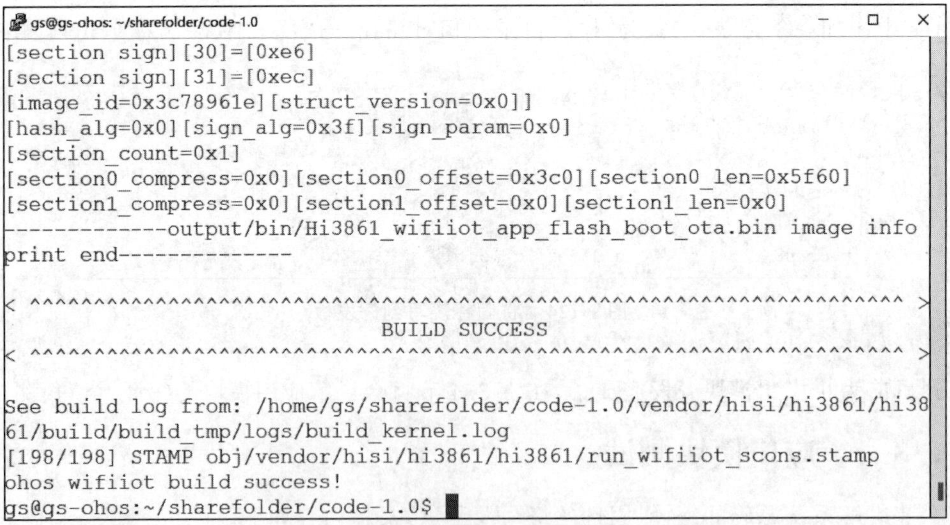

图 1-123 鸿蒙操作系统源码编译成功

2. 烧写固件

使用 HiBurn 将固件 Hi3861_wifiiot_app_allinone.bin 烧写到 Hi3861 开发板,固件的存放路径为\\out\wifiiot\Hi3861_wifiiot_app_allinone.bin,如图 1-124 所示。

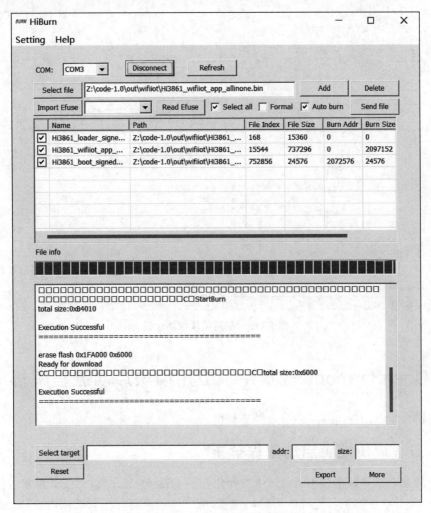

图 1-124 将固件烧写到开发板

注意:如果开发板连接失败,则可能是 COM 端口被 IPOP 终端工具占用,断开 IPOP 终端工具与开发板的连接即可。

3. 应用模块测试

运行 IPOP 终端工具,以便连接开发板,按复位键 RST 复位开发板,查看终端工具是否输出 Hello World 语句,如果输出就说明模块已经运行成功,运行效果如图 1-125 所示。

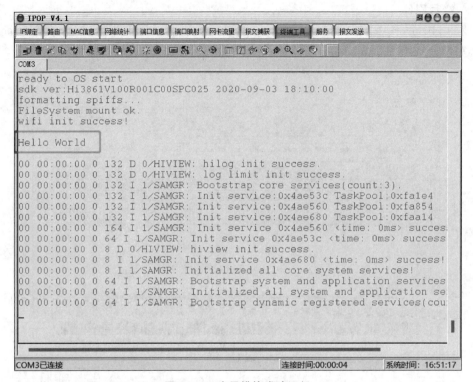

图 1-125　应用模块成功运行

1.4　OpenHarmony 轻量系统应用模块启动流程解析

本节讲解 OpenHarmony 轻量系统模块的启动流程及验证方法。

1.4.1　应用模块启动流程解析

由于鸿蒙轻量系统内核 LiteOS-M 被固化在了 Hi3861 芯片内，无法对鸿蒙内核进行修改及移植，并且内核在运行过程中无输出信息，所以启动流程分析只能从内核启动后的第 1 个入口函数 app_main 开始。应用模块的启动流程如图 1-126 所示。

函数 app_main 的源码位于 vendor/hisi/hi3861/hi3861/app/wifiiot_app/src/app_main.c 文件中，在 app_main 中实现了 OpenHarmony 启动初始化功能：

（1）获取并打印 SDK 版本号。
（2）外围设备初始化。
（3）工厂区 NV 初始化。
（4）Flash 分区表初始化。
（5）获取 Flash 分区表。

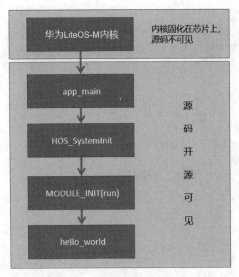

图 1-126　应用模块的启动流程

（6）非工厂区 NV 初始化。
（7）文件系统初始化。
（8）初始化事件资源。
（9）WiFi 初始化。
（10）最后调用 HOS_SystemInit 函数对鸿蒙操作系统进行初始化。

函数 app_main 的核心代码如下：

```
//vendor/hisi/hi3861/hi3861/app/wifiiot_app/src/app_main.c
hi_void app_main(hi_void)
{
    //获取 SDK 版本号
    const hi_char * sdk_ver = hi_get_sdk_version();
    printf("sdk ver:%s\r\n", sdk_ver);

    hi_flash_partition_table * ptable = HI_NULL;
    //外围设备初始化
    peripheral_init();
    peripheral_init_no_sleep();
    //工厂区 NV 初始化
    hi_u32 ret = hi_factory_nv_init(HI_FNV_DEFAULT_ADDR, HI_NV_DEFAULT_TOTAL_SIZE, HI_NV_DEFAULT_BLOCK_SIZE);
    if (ret != HI_ERR_SUCCESS) {
        printf("factory nv init fail\r\n");
    }
```

```c
    //Flash 分区表初始化
    ret = hi_flash_partition_init();
    if (ret != HI_ERR_SUCCESS) {
        printf("flash partition table init fail:0x%x \r\n", ret);
    }
    //获取 Flash 分区表
    ptable = hi_get_partition_table();
    //非工厂区 NV 初始化
    ret = hi_nv_init(ptable->table[HI_FLASH_PARTITON_NORMAL_NV].addr, ptable->table[HI_FLASH_PARTITON_NORMAL_NV].size,
        HI_NV_DEFAULT_BLOCK_SIZE);
    if (ret != HI_ERR_SUCCESS) {
        printf("nv init fail\r\n");
    }
    //文件系统初始化
    hi_fs_init();
    //初始化事件资源
    (hi_void)hi_event_init(APP_INIT_EVENT_NUM, HI_NULL);
    hi_sal_init();
    /* 如果此处设为 TRUE,则中断中的看门狗复位会显示复位时的 PC 值,但有复位不完全的风险,
因此量产版本务必设为 FALSE */
    hi_syserr_watchdog_Debug(HI_FALSE);
    /* 默认将宕机信息记录到 Flash,根据应用场景,可不记录,避免频繁异常情况降低 Flash
寿命 */
    hi_syserr_record_crash_info(HI_TRUE);
    hi_lpc_init();
    hi_lpc_register_hw_handler(config_before_sleep, config_after_sleep);
#if defined(CONFIG_AT_COMMAND) || defined(CONFIG_FACTORY_TEST_MODE)
    //初始化 AT 指令
    ret = hi_at_init();
    if (ret == HI_ERR_SUCCESS) {
        //注册系统 AT 指令
        hi_at_sys_cmd_register();
    }
#endif
    //初始化 WiFi
    ret = hi_wifi_init(APP_INIT_VAP_NUM, APP_INIT_USR_NUM);
    if (ret != HISI_OK) {
        printf("WiFi init failed!\n");
    } else {
        printf("WiFi init success!\n");
    }
    //OpenHarmony 系统初始化
    HOS_SystemInit();
}
```

函数 HOS_SystemInit 实现了对不同阶段的初始化,函数源码位于 base/startup/services/Bootstrap_lite/source/system_init.c 文件中,代码如下:

```c
//base/startup/services/Bootstrap_lite/source/system_init.c
void HOS_SystemInit(void)
{
  MODULE_INIT(bsp);
  MODULE_INIT(device);
  MODULE_INIT(core);
  SYS_INIT(service);
  SYS_INIT(feature);
  //初始化 run 段,源码中所有被宏 SYS_RUN 初始化的函数会在这里被调用执行
  MODULE_INIT(run);
  SAMGR_Bootstrap();
}
```

1.4.2 应用模块启动流程验证

本节验证 1.4.1 节的启动流程,整体思路如下:

(1) 在函数 app_main 中找到 HOS_SystemInit()函数,并在其上下行插入打印语句。

(2) 在函数 HOS_SystemInit()中找到 MODULE_INIT(run)语句,并在其上下行插入打印语句。

(3) 重新编译系统源码,烧写固件,查看打印消息。

在源码文件 vendor/hisi/hi3861/hi3861/app/wifiiot_app/src/app_main.c 中找到 HOS_SystemInit 函数,并在其上下行插入打印语句,代码如下:

```c
printf("%s %d\n", __FILE__, __LINE__);
HOS_SystemInit();
printf("%s %d\n", __FILE__, __LINE__);
```

插入效果如图 1-127 所示。

在源码文件 base/startup/services/Bootstrap_lite/source/system_init.c 中找到 MODULE_INIT(run)语句,并在其上下行插入打印语句,其中__FILE__用于获取当前程序执行的文件名称,__LINE__用于获取当前程序执行的行号,并引用 stdio.h 头文件,代码如下:

```c
printf("%s %d\n", __FILE__, __LINE__);
MODULE_INIT(run);
printf("%s %d\n", __FILE__, __LINE__);
```

插入效果如图 1-128 所示。

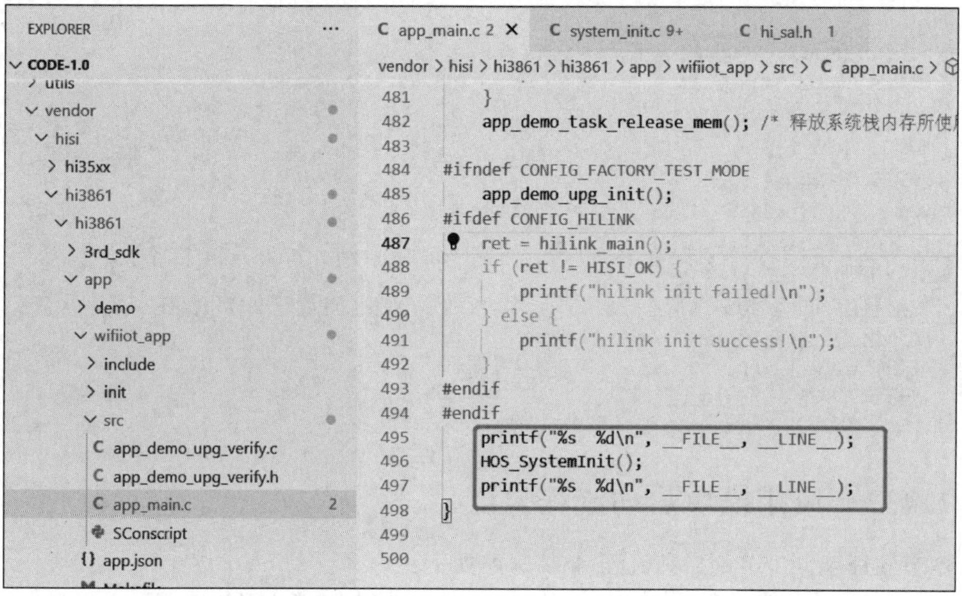

图 1-127 在函数 app_main 中插入打印语句

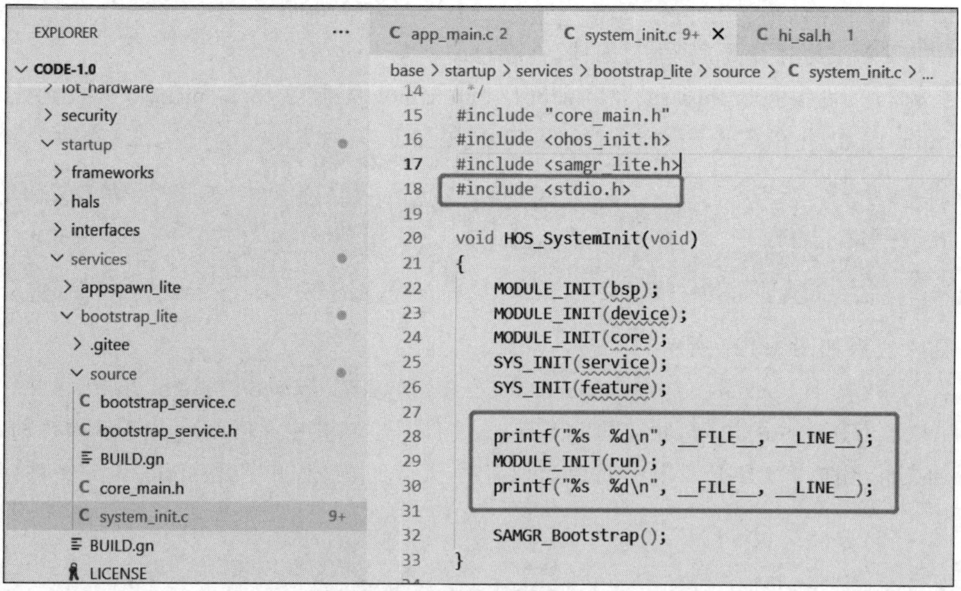

图 1-128 在函数 HOS_SystemInit 中插入打印语句

重新编译系统源码,烧写固件,查看打印消息,效果如图 1-129 所示。

```
ready to OS start
sdk ver:Hi3861V100R001C00SPC025 2020-09-03 18:10:00
formatting spiffs...
FileSystem mount ok.
wifi init success!
app/wifiiot_app/src/app_main.c   495
../../base/startup/services/bootstrap_lite/source/system_init.c   28

Hello World
../../base/startup/services/bootstrap_lite/source/system_init.c   30
app/wifiiot_app/src/app_main.c   497

00 00:00:00 0 132 D 0/HIVIEW: hilog init success.
00 00:00:00 0 132 D 0/HIVIEW: log limit init success.
00 00:00:00 0 132 I 1/SAMGR: Bootstrap core services(count:3).
00 00:00:00 0 132 I 1/SAMGR: Init service:0x4ae59c TaskPool:0xfa408
00 00:00:00 0 132 I 1/SAMGR: Init service:0x4ae608 TaskPool:0xfaa78
00 00:00:00 0 132 I 1/SAMGR: Init service:0x4ae728 TaskPool:0xfac38
```

图 1-129   应用模块启动流程验证

第 2 章 OpenHarmony 轻量系统设备开发

本章通过 25 个案例详细讲解 OpenHarmony 轻量系统核心设备接口 WatchDog、ADC、GPIO、PWM、I²C 开发技术。

2.1 案例 1：WatchDog

本案例讲解看门狗的使用，实现看门狗的启用、喂狗、关闭功能。

WatchDog 又叫看门狗，用于系统异常恢复，从本质上来讲就是一个定时器电路，一般有一个输入和一个输出，其中输入叫作喂狗，输出一般连接到另外一部分的复位端，一般连接到单片机。

WatchDog 的功能是定期地查看芯片内部的情况，一旦发生错误就向芯片发出重启信号。WatchDog 在程序的中断中拥有最高的优先级。

WatchDog 可应用在某些极端环境下（如真空环境、高温环境等），检查工作的单片机是否正常连续地工作，有关 WatchDog 的枚举、函数声明在头文件 wifiiot_watchdog.h 中。

Hi3861 芯片的 WatchDog 设备接口的开发流程如下：

(1) 分析应用模块的功能，确定在什么时候要对 WatchDog 进行操作。
(2) 调用函数 WatchDogEnable 启用 WatchDog（可选）。
(3) 调用函数 WatchDogKick 喂狗（可选）。
(4) 调用函数 WatchDogDisable 关闭看门狗（可选）。

函数 WatchDogEnable 的原型如下：

```
//base/iot_hardware/interfaces/kits/wifiiot_lite/wifiiot_watchdog.h
/**
 * @brief 启用看门狗
 *
 * @since 1.0
 * @version 1.0
 */
void WatchDogEnable(void);
```

函数 WatchDogKick 的原型如下：

```
//base/iot_hardware/interfaces/kits/wifiiot_lite/wifiiot_watchdog.h
/**
 * @brief 喂食看门狗
 *
 * @since 1.0
 * @version 1.0
 */
void WatchDogEnable(void);
```

函数 WatchDogDisable 的原型如下：

```
//base/iot_hardware/interfaces/kits/wifiiot_lite/wifiiot_watchdog.h
/**
 * @brief 禁用看门狗
 *
 * @since 1.0
 * @version 1.0
 */
void WatchDogDisable(void);
```

开发步骤如下：

（1）在 ohos_50 中创建本章目录 chapter_02，如图 2-1 所示。

（2）在 chapter_02 中创建工程目录 case1-watchdog，如图 2-2 所示。

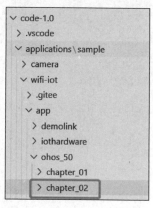

图 2-1　第 2 章工程集合目录

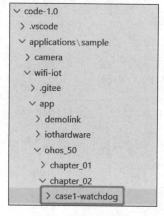

图 2-2　案例 1 的 WatchDog 工程目录

（3）在 case1-watchdog 中创建模块源码文件 watchdog_demo.c。

（4）引用必要的头文件，代码如下：

```c
//applications/sample/wifi-iot/app/ohos_50/chapter_02/case1-watchdog/watchdog_demo.c
#include <stdio.h>
#include <unistd.h>
#include <ohos_init.h>

#include "wifiiot_watchdog.h"
```

（5）创建函数 entry，实现看门狗启用、喂狗、关闭功能，代码如下：

```c
//applications/sample/wifi-iot/app/ohos_50/chapter_02/case1-watchdog/watchdog_demo.c
void entry(void)
{
    sleep(1);
    printf("WatchDog demo running!\n\r");
    //启用看门狗
    WatchDogEnable();
    printf("Enable watchdog\n");
    for (int i = 0; i < 3; i++)
    {
        //等待2s
        printf("Sleep 2 s\n");
        sleep(2);
        //喂狗
        WatchDogKick();
        printf("Kick the dog\n");
    }
    //关闭看门狗
    WatchDogDisable();
    printf("Disable watchdog\n");
}
```

（6）调用宏 APP_FEATURE_INIT，以便将模块入口函数初始化为 entry，代码如下：

```c
//将模块入口函数初始化为 entry
APP_FEATURE_INIT(entry);
```

（7）在 case1-watchdog 中创建模块，构建脚本 BUILD.gn 并配置模块，代码如下：

```
#//applications/sample/wifi-iot/app/ohos_50/chapter_02/case1-watchdog/BUILD.gn
#将生成的模块名称指定为 ch_02_watchdog
static_library("ch_02_watchdog") {
    #模块依赖的源码文件，当有多个文件时使用逗号隔开
```

```
    sources = [
        "watchdog_demo.c",
    ]
    #模块依赖的头文件存放路径,当有多个路径时使用逗号隔开
    include_dirs = [
        "//base/iot_hardware/interfaces/kits/wifiiot_lite",
    ]
}
```

(8) 将模块 ch_02_watchdog 配置到应用子系统,代码如下:

```
# //applications/sample/wifi-iot/app/BUILD.gn
import("//build/lite/config/component/lite_component.gni")

lite_component("app") {
    features = [
        "ohos_50/chapter_02/case1-watchdog:ch_02_watchdog",
    ]
}
```

(9) 测试:编译应用模块,将固件烧写到开发板,运行 IPOP 终端工具,以便与开发板相连,复位开发板。观察 IPOP 终端工具的打印信息,如图 2-3 所示。

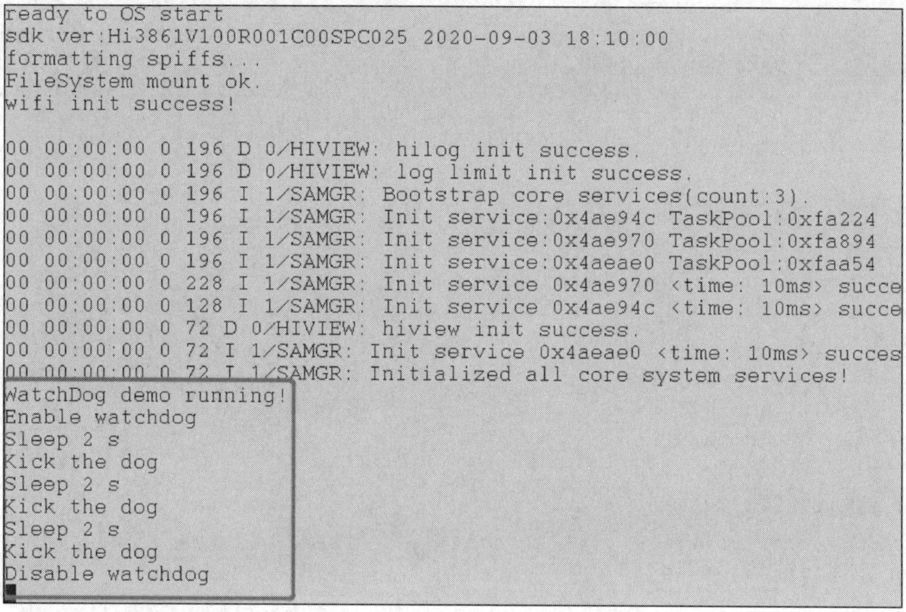

```
ready to OS start
sdk ver:Hi3861V100R001C00SPC025 2020-09-03 18:10:00
formatting spiffs...
FileSystem mount ok.
wifi init success!

00 00:00:00 0 196 D 0/HIVIEW: hilog init success.
00 00:00:00 0 196 D 0/HIVIEW: log limit init success.
00 00:00:00 0 196 I 1/SAMGR: Bootstrap core services(count:3).
00 00:00:00 0 196 I 1/SAMGR: Init service:0x4ae94c TaskPool:0xfa224
00 00:00:00 0 196 I 1/SAMGR: Init service:0x4ae970 TaskPool:0xfa894
00 00:00:00 0 196 I 1/SAMGR: Init service:0x4aeae0 TaskPool:0xfaa54
00 00:00:00 0 228 I 1/SAMGR: Init service 0x4ae970 <time: 10ms> succe
00 00:00:00 0 128 I 1/SAMGR: Init service 0x4ae94c <time: 10ms> succe
00 00:00:00 0 72 D 0/HIVIEW: hiview init success.
00 00:00:00 0 72 I 1/SAMGR: Init service 0x4aeae0 <time: 10ms> succes
00 00:00:00 0 72 I 1/SAMGR: Initialized all core system services!
WatchDog demo running!
Enable watchdog
Sleep 2 s
Kick the dog
Sleep 2 s
Kick the dog
Sleep 2 s
Kick the dog
Disable watchdog
```

图 2-3 WatchDog 的运行效果

2.2 ADC

模数转换器（Analog-to-Digital Converter，ADC）实现对外部模拟信号转换成一定比例的数字值，从而实现对模拟信号的测量，可应用于电量、雨量、土壤湿度、光照强度等检测，有关 ADC 的枚举、函数声明在头文件 wifiiot_adc.h 中。

Hi3861 芯片的 ADC 设备接口的开发流程如下：

(1) 分析电路，确定 ADC 通道索引。

(2) 调用函数 AdcRead 读取模拟数据。

函数 AdcRead 的原型如下：

```
//base/iot_hardware/interfaces/kits/wifiiot_lite/wifiiot_adc.h
/**
 * @brief 基于输入参数从指定的 ADC 通道读取一段采样数据
 * @param channel 为 wifiiotAdcChannelIndex 枚举类型，表示 ADC 通道索引
 * @param data 为 unsigned short 类型变量的指针，用来保存读取到的模拟数据
 * @param equModel 为 wifiiotAdcEquModelSel 枚举类型，表示方程模型
 * @param curBais 为 wifiiotAdcCurBais 枚举类型，表示模拟功率控制模式
 * @param rstCnt 表示从重置到转换开始的时间计数。一个计数等于 334ns。该值必须在 0 到 0xFF
 * @return 如果操作成功,则返回{@link WIFI_IOT_SUCCESS},如果为否,则返回{@link wifiiot_
 * errno.h}中定义的错误代码
 */
unsigned int AdcRead(wifiiotAdcChannelIndex channel, unsigned short * data, wifiiotAdcEquModelSel
equModel,wifiiotAdcCurBais curBais, unsigned short rstCnt);
```

枚举 wifiiotAdcChannelIndex 的原型如下：

```
//base/iot_hardware/interfaces/kits/wifiiot_lite/wifiiot_adc.h
/**
 * @brief Enumerates ADC channel indexes.
 *
 */
typedef enum {
    /** Channel 0 */
    WIFI_IOT_ADC_CHANNEL_0,
    /** Channel 1 */
    WIFI_IOT_ADC_CHANNEL_1,
    /** Channel 2 */
    WIFI_IOT_ADC_CHANNEL_2,
    /** Channel 3 */
    WIFI_IOT_ADC_CHANNEL_3,
    /** Channel 4 */
```

```
    WIFI_IOT_ADC_CHANNEL_4,
    /** Channel 5 */
    WIFI_IOT_ADC_CHANNEL_5,
    /** Channel 6 */
    WIFI_IOT_ADC_CHANNEL_6,
    /** Channel 7 */
    WIFI_IOT_ADC_CHANNEL_7,
    /** Button value */
    WIFI_IOT_ADC_CHANNEL_BUTT,
} wifiiotAdcChannelIndex;
```

枚举 wifiiotAdcEquModelSel 的原型如下：

```
//base/iot_hardware/interfaces/kits/wifiiot_lite/wifiiot_adc.h
/**
 * @brief Enumerates analog power control modes.
 */
typedef enum {
    /** Automatic control */
    WIFI_IOT_ADC_CUR_BAIS_DEFAULT,
    /** Automatic control */
    WIFI_IOT_ADC_CUR_BAIS_AUTO,
    /** Manual control (AVDD = 1.8 V) */
    WIFI_IOT_ADC_CUR_BAIS_1P8V,
    /** Manual control (AVDD = 3.3 V) */
    WIFI_IOT_ADC_CUR_BAIS_3P3V,
    /** Button value */
    WIFI_IOT_ADC_CUR_BAIS_BUTT,
} wifiiotAdcCurBais;
```

枚举 wifiiotAdcCurBais 的原型如下：

```
//base/iot_hardware/interfaces/kits/wifiiot_lite/wifiiot_adc.h
/**
 * @brief Enumerates equation models.
 */
typedef enum {
    /** One-equation model */
    WIFI_IOT_ADC_EQU_MODEL_1,
    /** Two-equation model */
    WIFI_IOT_ADC_EQU_MODEL_2,
    /** Four-equation model */
    WIFI_IOT_ADC_EQU_MODEL_4,
    /** Eight-equation model */
    WIFI_IOT_ADC_EQU_MODEL_8,
    /** Button value */
    WIFI_IOT_ADC_EQU_MODEL_BUTT,
} wifiiotAdcEquModelSel;
```

2.2.1 案例2：雨滴探测器

本案例讲解雨滴探测器,实现天气状态的监测功能,其中核心模块为雨滴探测模块,如图 2-4 所示。

雨滴探测模块可以测量雨量,对各种天气状况都可以进行监测,并转换成数字信号和 AO 输出。

图 2-4 中左边为雨滴探测模块,右边为信号处理模块,雨滴探测模块与信号处理模块通过两根母对母的杜邦线连接。信号处理模块下面的 4 个引脚从左往右依次为 AO(模拟输出)、DO(数字输出)、GND、5V VCC。除 DO 外分别与主控板的 GPIO12(ADC0)、GND、+5V 相连,其中 DO 不与主控板相连。

雨滴探测器是根据雨滴探测模块 ADC0 通道输出的模拟量值判断当前雨滴的多少。

ADC0 为雨滴探测模拟信号的输出端,当 ADC0 输出的模拟值在 1200 以上时表示无水滴,在 1200~1000 时表示有少量水滴,在 1000 以下时表示有大量水滴。

开发步骤如下:

(1) 在 chapter_02 中创建雨滴探测器工程目录 case2-water_drop_detector,如图 2-5 所示。

图 2-4 雨滴探测模块

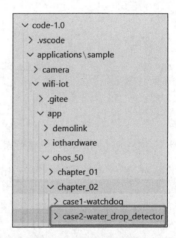

图 2-5 案例 2 的雨滴探测器工程目录

(2) 在 case2-water_drop_detector 中创建源码文件 water_drop_detector_demo.c。

(3) 引用必要的头文件、声明 ADC_CHANNEL 宏、声明 unsigned short 类型变量 data,代码如下:

```
//applications/sample/wifi-iot/app/ohos_50/chapter_02/case2-water_drop_detector/water_drop_detector_demo.c
#include <stdio.h>
#include <unistd.h>
```

```
#include <ohos_init.h>
#include <wifiiot_adc.h>
//将使用的 ADC CHANNEL 序号声明为 0
#define ADC_CHANNEL WIFI_IOT_ADC_CHANNEL_0
//用来保存读取到的模拟数据
unsigned short data = 0;
```

(4) 创建函数 entry，通过 ADC0 通道循环读取当前雨滴的模拟数据，实现打印输出当前天气状态的功能，代码如下：

```
//applications/sample/wifi-iot/app/ohos_50/chapter_02/case2-water_drop_detector/water_drop_detector_demo.c
void entry(void)
{
    //等待系统初始化
    sleep(1);
    while (1)
    {
        /**
         * 调用函数 AdcRead 读取模拟数据，参数说明如下
         *    - ADC_CHANNEL 表示 ADC 通道索引
         *    - &data 表示指向存储读取数据的地址的指针
         *    - WIFI_IOT_ADC_EQU_MODEL_4 表示方程模型
         *    - WIFI_IOT_ADC_CUR_BAIS_DEFAULT 表示模拟功率控制模式
         *    - 0 表示从重置到转换开始的时间计数(一个计数等于 334 ns)
         */
        AdcRead(ADC_CHANNEL, &data, WIFI_IOT_ADC_EQU_MODEL_4, WIFI_IOT_ADC_CUR_BAIS_DEFAULT, 0);
        printf("Rainfall volume:%d\t",data);
        if(data>1200){
            printf("no rain\r\n");
        }else if (data>1000)
        {
            printf("light rain\r\n");
        }else{
            printf("heavy rain\r\n");
        }
        sleep(1);
    }
}
```

(5) 调用宏 APP_FEATURE_INIT 并将模块入口函数初始化为 entry，代码如下：

```
//将模块入口函数初始化为 entry
APP_FEATURE_INIT(entry);
```

(6) 在 case2-water_drop_detector 中创建模块，构建脚本 BUILD.gn 并配置模块，代码如下：

```
#//applications/sample/wifi-iot/app/ohos_50/chapter_02/case2-water_drop_detector/BUILD.gn
#将生成的模块名称指定为 ch_02_water_drop_detector
static_library("ch_02_water_drop_detector") {
    #模块依赖的源码文件,当有多个文件时使用逗号隔开
    sources = [
        "water_drop_detector_demo.c",
    ]
    #模块依赖的头文件存放路径,当有多个路径时使用逗号隔开
    include_dirs = [
        "//base/iot_hardware/interfaces/kits/wifiiot_lite",
    ]
}
```

(7) 将模块 ch_02_water_drop_detector 配置到应用子系统，代码如下：

```
#//applications/sample/wifi-iot/app/BUILD.gn
import("//build/lite/config/component/lite_component.gni")
lite_component("app") {
    features = [
        "case2-water_drop_detector:ch_02_water_drop_detector",
    ]
}
```

(8) 测试：编译应用模块，将固件烧写到开发板，将雨滴探测模块与开发板相连，运行 IPOP 终端工具以便与开发板相连，复位开发板。改变雨滴探测模块上的水量，观察 IPOP 终端工具的打印信息，如图 2-6 所示。

```
00 00:00:00 0 196 I 1/SAMGR: Init service:0x4ae8ac TaskPool:0xfa224
00 00:00:00 0 196 I 1/SAMGR: Init service:0x4ae8d0 TaskPool:0xfa894
00 00:00:00 0 196 I 1/SAMGR: Init service:0x4aea20 TaskPool:0xfaa54
00 00:00:00 0 228 I 1/SAMGR: Init service 0x4ae8d0 <time: 0ms> succes
00 00:00:00 0 128 I 1/SAMGR: Init service 0x4ae8ac <time: 0ms> succes
00 00:00:00 0 72 D 0/HIVIEW: hiview init success.
00 00:00:00 0 72 I 1/SAMGR: Init service 0x4aea20 <time: 0ms> success
00 00:00:00 0 72 I 1/SAMGR: Initialized all core system services!
Rainfall volume:2099    no rain
Rainfall volume:1172    light rain
Rainfall volume:1034    light rain
Rainfall volume:960     heavy rain
Rainfall volume:972     heavy rain
Rainfall volume:2073    no rain
Rainfall volume:612     heavy rain
Rainfall volume:2068    no rain
Rainfall volume:899     heavy rain
Rainfall volume:873     heavy rain
Rainfall volume:816     heavy rain
Rainfall volume:807     heavy rain
Rainfall volume:2032    no rain
Rainfall volume:2040    no rain
Rainfall volume:2036    no rain
Rainfall volume:2034    no rain
Rainfall volume:2030    no rain
```

图 2-6　雨滴探测器的运行效果

2.2.2 案例3：游戏杆

本案例讲解游戏杆，实现游戏杆改变游戏人物的位置坐标功能，核心模块为双轴按键摇杆模块，如图 2-7 所示。

双轴按键摇杆传感器模块采用 PS2 摇杆电位器制作，具有 2 轴(X,Y)模拟输出，1 路(Z)按钮数字输出。

图 2-7 中左面 5 个引脚从上往下依次为 GND、+5V、URX(模拟输出)、URY(模拟输出)、SW(数字输出)。除 SW 外分别与主控板的 GND、+5V、GPIO04(ADC1)、GPIO05(ADC2)相连，其中 SW 不与主控板相连。

案例实现原理：当游戏杆的位置发生改变后，通过两路 ADC 分别读取操纵杆当前 x 轴和 y 轴的坐标值发生的变化，游戏中人物对应的 x 轴和 y 轴的坐标值也相应地发生改变。

开发步骤如下：

(1) 在 chapter_02 中创建工程目录 case3-joystick，如图 2-8 所示。

图 2-7 双轴按键摇杆模块

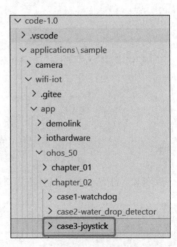

图 2-8 案例 3 的游戏杆工程目录

(2) 在 case3-joystick 中创建源码文件 joystick_demo.c。

(3) 引用必要的头文件、声明所需的宏及全局变量，代码如下：

```
//applications/sample/wifi-iot/app/ohos_50/chapter_02/case3-joystick/joystick_demo.c
#include <stdio.h>
#include <unistd.h>
#include <ohos_init.h>
#include <wifiiot_adc.h>
//将使用的 ADC CHANNEL 序号声明为 1 与 2
#define ADC_CHANNEL_X WIFI_IOT_ADC_CHANNEL_1
#define ADC_CHANNEL_Y WIFI_IOT_ADC_CHANNEL_2
```

```c
//用来保存读取到的模拟数据
unsigned short data_x = 0;
unsigned short data_y = 0;
```

(4) 创建函数 entry, 实现循环读取并打印 ADC1、ADC2 引脚的模拟数据功能, 代码如下:

```c
//applications/sample/wifi-iot/app/ohos_50/chapter_02/case3-joystick/joystick_demo.c
void entry(void)
{
    sleep(1);
    while (1)
    {
        //读取 WIFI_IOT_ADC_CHANNEL_1 模拟数据
        AdcRead(ADC_CHANNEL_X, &data_x, WIFI_IOT_ADC_EQU_MODEL_2, WIFI_IOT_ADC_CUR_BAIS_DEFAULT, 0);
        usleep(1000);
        //读取 WIFI_IOT_ADC_CHANNEL_2 模拟数据
        AdcRead(ADC_CHANNEL_Y, &data_y, WIFI_IOT_ADC_EQU_MODEL_2, WIFI_IOT_ADC_CUR_BAIS_DEFAULT, 0);
        printf("X = %d,Y = %d\n", data_x,data_y);
        sleep(1);
    }
}
```

(5) 调用宏 APP_FEATURE_INIT 并将模块入口函数初始化为 entry, 代码如下:

```c
//将模块入口函数初始化为 entry
APP_FEATURE_INIT(entry);
```

(6) 在 case3-joystick 中创建模块, 构建脚本 BUILD.gn 并配置模块, 代码如下:

```
#//applications/sample/wifi-iot/app/ohos_50/chapter_02/case3-joystick/BUILD.gn
static_library("ch_02_joystick") {
    sources = [
        "joystick_demo.c",
    ]
    include_dirs = [
        "//base/iot_hardware/interfaces/kits/wifiiot_lite",
    ]
}
```

(7) 将模块 ch_02_joystick 配置到应用子系统,代码如下:

```
#//applications/sample/wifi-iot/app/BUILD.gn
import("//build/lite/config/component/lite_component.gni")
lite_component("app") {
    features = [
        "ohos_50/chapter_02/case3-joystick:ch_02_joystick",
    ]
}
```

(8) 测试:编译应用模块,将固件烧写到开发板,运行 IPOP 终端工具,以便与开发板相连,复位开发板。推动操纵杆,观察 IPOP 终端工具打印的坐标信息,如图 2-9 所示。

```
ready to OS start
sdk ver:Hi3861V100R001C00SPC025 2020-09-03 18:10:00
FileSystem mount ok.
wifi init success!

00 00:00:00 0 132 D 0/HIVIEW: hilog init success.
00 00:00:00 0 132 D 0/HIVIEW: log limit init success.
00 00:00:00 0 132 I 1/SAMGR: Bootstrap core services(count:3).
00 00:00:00 0 132 I 1/SAMGR: Init service:0x4ae63c TaskPool:0xfa1e4
00 00:00:00 0 132 I 1/SAMGR: Init service:0x4ae660 TaskPool:0xfa854
00 00:00:00 0 132 I 1/SAMGR: Init service:0x4ae77c TaskPool:0xfaa14
00 00:00:00 0 164 I 1/SAMGR: Init service 0x4ae660 <time: 0ms> success
00 00:00:00 0 64 I 1/SAMGR: Init service 0x4ae63c <time: 0ms> success
00 00:00:00 0 8 D 0/HIVIEW: hiview init success.
00 00:00:00 0 8 I 1/SAMGR: Init service 0x4ae77c <time: 0ms> success!
00 00:00:00 0 8 I 1/SAMGR: Initialized all core system services!
X=1854,Y=1352
X=1852,Y=1352
X=2074,Y=1366
```

图 2-9 游戏杆的运行效果

注意:如果出现乱码,则可能是 ADC 通道存在问题,更换 ADC 通道即可。

2.2.3 案例 4:烟雾探测器

本案例讲解烟雾探测器,实现空气中的烟雾浓度监测功能,其中核心模块为 MQ-2 型烟雾传感器模块,如图 2-10 所示。

MQ-2 型烟雾传感器是用二氧化锡半导体气敏材料制成的传感器,属于表面离子式 N 型半导体。处于 200~300℃时,二氧化锡吸附空气中的氧,形成氧的负离子吸附,使半导体中的电子密度减小,从而使其电阻值增加。当与烟雾接触时,如果晶粒间界处的势垒收到烟雾的浓度而变化,就会引起表面导电率的变化。利用这一点就可以获得这种烟雾存

图 2-10 MQ-2 型烟雾传感器模块

在的信息,烟雾的浓度越大,导电率越大,输出电阻越小,输出的模拟信号就越大。

4个引脚分别为VCC、GND、DO、AO。除DO外分别与主控板的+5V、GND、GPIO07(ADC3)相连。

实现原理:通过ADC三通道读取到MQ-2型烟雾传感器输出的烟雾浓度值。

开发步骤如下:

(1) 在chapter_02中创建工程目录case3-smoke_detector。

(2) 在case3-smoke_detector中创建源码文件smoke_detector_demo.c。

(3) 引用必要的头文件、声明所需的宏及全局变量,代码如下:

```c
//applications/sample/wifi-iot/app/ohos_50/chapter_02/case3-smoke_detector/smoke_detector_demo.c
#include <stdio.h>
#include <unistd.h>
#include <ohos_init.h>
#include <wifiiot_adc.h>
//将使用的ADC CHANNEL序号声明为3
#define ADC_MQ2 WIFI_IOT_ADC_CHANNEL_3
//用来保存读取到的模拟数据
unsigned short data = 0;
```

(4) 创建函数entry,实现循环读取并打印ADC3引脚的模拟数据功能,代码如下:

```c
//applications/sample/wifi-iot/app/ohos_50/chapter_02/case3-smoke_detector/smoke_detector_demo.c
void entry(void)
{
    sleep(1);
    while (1)
    {
        AdcRead(ADC_MQ2, &data, WIFI_IOT_ADC_EQU_MODEL_2, WIFI_IOT_ADC_CUR_BAIS_DEFAULT, 0);
        printf("mq2 value:%d\n", data);
        sleep(1);
    }
}
```

(5) 调用宏APP_FEATURE_INIT并将模块入口函数初始化为entry,代码如下:

```c
//将模块入口函数初始化为entry
APP_FEATURE_INIT(entry);
```

(6) 在case3-smoke_detector中创建模块,构建脚本BUILD.gn并初始化模块,代码如下:

```
#//applications/sample/wifi-iot/app/ohos_50/chapter_02/case3-smoke_detector/BUILD.gn
static_library("ch_02_smoke_detector") {
    sources = [
        "smoke_detector_demo.c",
    ]
    include_dirs = [
        "//base/iot_hardware/interfaces/kits/wifiiot_lite",
    ]
}
```

（7）将模块 ch_02_smoke_detector 配置到应用子系统，代码如下：

```
#//applications/sample/wifi-iot/app/BUILD.gn
import("//build/lite/config/component/lite_component.gni")
lite_component("app") {
    features = [
        "ohos_50/chapter_02/case3-smoke_detector:ch_02_smoke_detector",
    ]
}
```

（8）测试：编译应用模块，将固件烧写到开发板，将 MQ-2 型烟雾传感器模块与开发板相连，运行 IPOP 终端工具，以便与开发板相连，复位开发板，使用打火机在 MQ-2 型烟雾传感器探头处放气。观察 IPOP 终端工具的打印信息，如图 2-11 所示。

```
00 00:00:00 0 196 D 0/HIVIEW: hilog init success.
00 00:00:00 0 196 D 0/HIVIEW: log limit init success.
00 00:00:00 0 196 I 1/SAMGR: Bootstrap core services(count:3).
00 00:00:00 0 196 I 1/SAMGR: Init service:0x4ae8ac TaskPool:0xfa224
00 00:00:00 0 196 I 1/SAMGR: Init service:0x4ae8d0 TaskPool:0xfa894
00 00:00:00 0 196 I 1/SAMGR: Init service:0x4aea5c TaskPool:0xfaa54
00 00:00:00 0 228 I 1/SAMGR: Init service 0x4ae8d0 <time: 0ms> succes
00 00:00:00 0 128 I 1/SAMGR: Init service 0x4ae8ac <time: 0ms> succes
00 00:00:00 0 72 D 0/HIVIEW: hiview init success.
00 00:00:00 0 72 I 1/SAMGR: Init service 0x4aea5c <time: 0ms> success
00 00:00:00 0 72 I 1/SAMGR: Initialized all core system services!
mq2 value:276
Smoke concentration is normal.
mq2 value:276
Smoke concentration is normal.
mq2 value:278
Smoke concentration is normal.
mq2 value:280
Smoke concentration is normal.
mq2 value:1662
*Warning! Smoke concentration is too high.Please ventilate in time.
mq2 value:2112
*Warning! Smoke concentration is too high.Please ventilate in time.
```

图 2-11　烟雾探测器的运行效果

注意：在测试时，需要先让 MQ-2 型烟雾传感器运行一段时间，再用打火机放气，否则可能采集到的值变化不明显。

2.2.4 案例5：声音监测仪

本案例讲解声音监测仪，实现环境声音响度的监测功能，核心模块为高感度话筒传感器模块，如图2-12所示。

图 2-12 高感度话筒传感器模块

该模块由一个用于检测声音的灵敏电容话筒和一个放大电路组成。该模块的输出既是模拟的又是数字的，并且当声音强度达到某个阈值时激活。灵敏度阈值可以通过传感器上的电位计进行调整。模拟输出电压会随话筒接收的声音强度而变化。

图2-12中右面圆柱形器件为灵敏电容话筒，左面的4个引脚分别为AO、G、＋、DO，除DO外分别与主控板的GPIO09(ADC4)、GND、＋5V相连。

开发步骤如下：

(1) 在chapter_02中创建工程目录case5-noise_detector。

(2) 在case5-noise_detector中创建源码文件noise_detector_demo.c。

(3) 功能实现，代码如下：

```c
//applications/sample/wifi-iot/app/ohos_50/chapter_02/case5-noise_detector/noise_detector_demo.c
#include <stdio.h>
#include <unistd.h>
#include <ohos_init.h>
#include <wifiiot_adc.h>
//将使用的ADC CHANNEL序号声明为0
#define ADC_MQ2 WIFI_IOT_ADC_CHANNEL_4
//用来保存读取到的模拟数据
unsigned short data = 0;
void entry(void)
{
    sleep(1);
    while (1)
    {
        AdcRead(ADC_MQ2, &data, WIFI_IOT_ADC_EQU_MODEL_2, WIFI_IOT_ADC_CUR_BAIS_DEFAULT, 0);
        printf("noise value:%d\r", data);
        usleep(100000);
    }
}
APP_FEATURE_INIT(entry);
```

(4) 在 case5-noise_detector 中创建模块,构建脚本 BUILD.gn 并初始化模块,代码如下:

```
#//applications/sample/wifi-iot/app/ohos_50/chapter_02/case5-noise_detector/BUILD.gn
static_library("ch_02_noise_detector") {
    sources = [
        "noise_detector_demo.c",
    ]
    include_dirs = [
        "//base/iot_hardware/interfaces/kits/wifiiot_lite",
    ]
}
```

(5) 将模块 ch_02_noise_detector 配置到应用子系统,代码如下:

```
#//applications/sample/wifi-iot/app/BUILD.gn
import("//build/lite/config/component/lite_component.gni")
lite_component("app") {
    features = [
        "ohos_50/chapter_02/case5-noise_detector:ch_02_noise_detector",
    ]
}
```

(6) 测试:编译应用模块,将固件烧写到开发板,运行 IPOP 终端工具,以便与开发板相连,复位开发板。观察 IPOP 终端工具的打印信息,如图 2-13 所示。

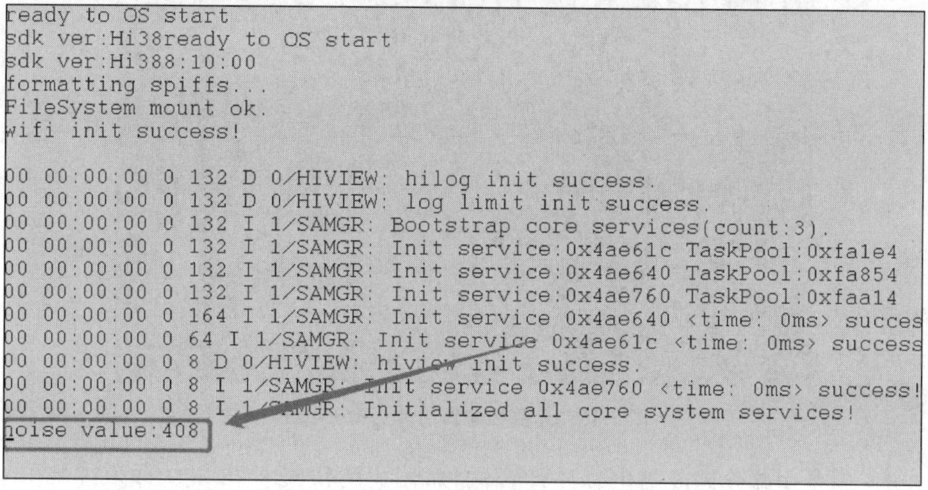

```
ready to OS start
sdk ver:Hi38ready to OS start
sdk ver:Hi388:10:00
formatting spiffs...
FileSystem mount ok.
wifi init success!

00 00:00:00 0 132 D 0/HIVIEW: hilog init success.
00 00:00:00 0 132 D 0/HIVIEW: log limit init success.
00 00:00:00 0 132 I 1/SAMGR: Bootstrap core services(count:3).
00 00:00:00 0 132 I 1/SAMGR: Init service:0x4ae61c TaskPool:0xfa1e4
00 00:00:00 0 132 I 1/SAMGR: Init service:0x4ae640 TaskPool:0xfa854
00 00:00:00 0 132 I 1/SAMGR: Init service:0x4ae760 TaskPool:0xfaa14
00 00:00:00 0 164 I 1/SAMGR: Init service 0x4ae640 <time: 0ms> succes
00 00:00:00 0 64 I 1/SAMGR: Init service 0x4ae61c <time: 0ms> success
00 00:00:00 0 8 D 0/HIVIEW: hiview init success.
00 00:00:00 0 8 I 1/SAMGR: Init service 0x4ae760 <time: 0ms> success!
00 00:00:00 0 8 I 1/SAMGR: Initialized all core system services!
noise value:408
```

图 2-13 噪声检测仪的运行效果

2.2.5 案例 6：光照检测仪

本案例讲解光照检测仪，实现光照强度采集功能，其中核心模块为光敏电阻，如图 2-14 所示。

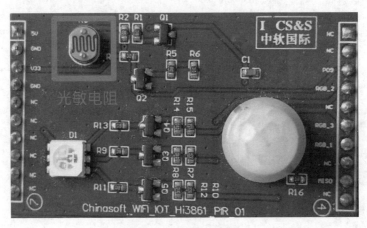

图 2-14 光敏电阻模块

图 2-14 中左面从上数第 2 个和第 3 个引脚分别为 V33(3.3V 电源)、GND(地线)。右面从上数第 3 个引脚为 P09(GPIO09，也就是 ADC4)。将主控板插到母板上，再将人体红外感应板插到母板上，如图 2-15 所示。

图 2-15 光照检测仪组装后的开发板

光敏电阻是用硫化镉或硒化镉等半导体材料制成的特殊电阻，其工作原理基于光电效应。光照越强，阻值就越低，随着光照强度的升高，电阻值迅速降低，亮电阻值可小至 1kΩ 以下。光敏电阻对光线十分敏感，其在无光照时，呈高阻状态，暗电阻一般可达 1.5MΩ。

开发步骤如下：

(1) 在 chapter_02 中创建工程目录 case6-light_detector。

(2) 在 case6-light_detector 中创建源码文件 light_detector_demo.c。

(3) 功能实现，代码如下：

```c
//applications/sample/wifi-iot/app/ohos_50/chapter_02/case6-light_detector/light_detector_demo.c
#include <stdio.h>
#include <unistd.h>
#include <ohos_init.h>
#include <wifiiot_adc.h>
//将使用的 ADC CHANNEL 序号声明为 4
#define ADC_LIGHT WIFI_IOT_ADC_CHANNEL_4
//用来保存读取到的模拟数据
unsigned short data = 0;
void entry(void)
{
    sleep(1);
    while (1)
    {
        AdcRead(ADC_LIGHT, &data, WIFI_IOT_ADC_EQU_MODEL_4, WIFI_IOT_ADC_CUR_BAIS_DEFAULT, 0);
        printf("light value:%d\n", data);
        sleep(1);
    }
}
APP_FEATURE_INIT(entry);
```

(4) 在 case6-light_detector 中创建模块，构建脚本 BUILD.gn 并初始化模块，代码如下：

```
#//applications/sample/wifi-iot/app/ohos_50/chapter_02/case6-light_detector/BUILD.gn
static_library("ch_02_light_detector") {
    sources = [
        "light_detector_demo.c",
    ]
    include_dirs = [
        "//base/iot_hardware/interfaces/kits/wifiiot_lite",
    ]
}
```

(5) 将模块 ch_02_light_detector 配置到应用子系统，代码如下：

```
#//applications/sample/wifi-iot/app/BUILD.gn
import("//build/lite/config/component/lite_component.gni")
lite_component("app") {
```

```
    features = [
        "ohos_50/chapter_02/case6-light_detector:ch_02_light_detector",
    ]
}
```

(6)测试:编译应用模块,将固件烧写到开发板,运行 IPOP 终端工具,以便与开发板相连,复位开发板。观察 IPOP 终端工具的打印信息,如图 2-16 所示。

```
ready to OS start
sdk ver:Hi3861V100R001C00SPC025 2020-09-03 18:10:00
formatting spiffs...
FileSystem mount ok.
wifi init success!

00 00:00:00 0 132 D 0/HIVIEW: hilog init success.
00 00:00:00 0 132 D 0/HIVIEW: log limit init success.
00 00:00:00 0 132 I 1/SAMGR: Bootstrap core services(count:3).
00 00:00:00 0 132 I 1/SAMGR: Init service:0x4ae5bc TaskPool:0xfa1e4
00 00:00:00 0 132 I 1/SAMGR: Init service:0x4ae5e0 TaskPool:0xfa854
00 00:00:00 0 132 I 1/SAMGR: Init service:0x4ae700 TaskPool:0xfaa14
00 00:00:00 0 164 I 1/SAMGR: Init service 0x4ae5e0 <time: 0ms> success
00 00:00:00 0 64 I 1/SAMGR: Init service 0x4ae5bc <time: 0ms> success
00 00:00:00 0 8 D 0/HIVIEW: hiview init success.
00 00:00:00 0 8 I 1/SAMGR: Init service 0x4ae700 <time: 0ms> success!
00 00:00:00 0 8 I 1/SAMGR: Initialized all core system services!
light value:132
light value:132
light value:132
light value:140
light value:1841
light value:132
light value:132
```

图 2-16 光照检测仪的运行效果

2.2.6 案例 7:生命探测仪

本案例讲解生命探测仪,实现生命探测功能,其中核心模块为人体红外传感器 AS312,如图 2-17 所示。

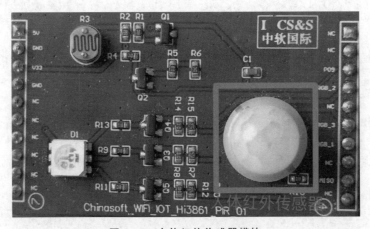

图 2-17 人体红外传感器模块

右面从上往下数第 9 个引脚为 P07(GPIO07,也就是 ADC3)。将主控板插到母板上,再将人体红外感应板插到母板上,如图 2-18 所示。

图 2-18　生命探测仪组装后的开发板

AS312 是将数字智能控制电路与人体探测敏感元都集成在电磁屏蔽罩内的热释电红外传感器。人体探测敏感元将感应到的人体移动信号通过甚高阻抗差分输入电路耦合到数字智能集成电路芯片上,数字智能集成电路将信号转化成 15 位的 ADC 数字信号,当 PIR 信号超过选定的数字阈值时就会有延时的 REL 电平输出。所有的信号处理都在芯片上完成。

开发步骤如下:

(1) 在 chapter_02 中创建工程目录 case7-life_detector。

(2) 在 case7-life_detector 中创建源码文件 life_detector_demo.c。

(3) 功能实现,代码如下:

```
//applications/sample/wifi-iot/app/ohos_50/chapter_02/case7-life_detector/life_detector_demo.c
#include <stdio.h>
#include <unistd.h>
#include <ohos_init.h>
#include <wifiiot_adc.h>
//将使用的 ADC CHANNEL 序号声明为 3
#define ADC_LIGHT WIFI_IOT_ADC_CHANNEL_3
//用来保存读取到的模拟数据
unsigned short data = 0;
```

```c
void entry(void)
{
    sleep(1);
    while (1)
    {
        AdcRead(ADC_LIGHT, &data, WIFI_IOT_ADC_EQU_MODEL_4, WIFI_IOT_ADC_CUR_BAIS_DEFAULT, 0);
        printf("life value: % d\n", data);
        usleep(300000);
    }
}
APP_FEATURE_INIT(entry);
```

(4) 在 case7-life_detector 中创建模块，构建脚本 BUILD.gn 并初始化模块，代码如下：

```
#//applications/sample/wifi-iot/app/ohos_50/chapter_02/case7-life_detector/BUILD.gn
static_library("ch_02_life_detector") {
    sources = [
        "life_detector_demo.c",
    ]
    include_dirs = [
        "//base/iot_hardware/interfaces/kits/wifiiot_lite",
    ]
}
```

(5) 将模块 ch_02_life_detector 配置到应用子系统，代码如下：

```
#//applications/sample/wifi-iot/app/BUILD.gn
import("//build/lite/config/component/lite_component.gni")
lite_component("app") {
    features = [
        "ohos_50/chapter_02/case7-life_detector:ch_02_life_detector",
    ]
}
```

(6) 测试：编译应用模块，将固件烧写到开发板，运行 IPOP 终端工具，以便与开发板相连，复位开发板，在人体生命传感器上挥手。观察 IPOP 终端工具的打印信息，如图 2-19 所示。

```
ready to OS start
sdk ver:Hi3861V100R001C00SPC025 2020-09-03 18:10:00
FileSystem mount ok.
wifi init success!
00 00:00:00 0 132 D 0/HIVIEW: hilog init success.
00 00:00:00 0 132 D 0/HIVIEW: log limit init success.
00 00:00:00 0 132 I 1/SAMGR: Bootstrap core services(count:3).
00 00:00:00 0 132 I 1/SAMGR: Init service:0x4ae61c TaskPool:0xfa1e4
00 00:00:00 0 132 I 1/SAMGR: Init service:0x4ae640 TaskPool:0xfa854
00 00:00:00 0 132 I 1/SAMGR: Init service:0x4ae760 TaskPool:0xfaa14
00 00:00:00 0 164 I 1/SAMGR: Init service 0x4ae640 <time: 0ms> success
00 00:00:00 0 64 I 1/SAMGR: Init service 0x4ae61c <time: 0ms> success
00 00:00:00 0 8 D 0/HIVIEW: hiview init success.
00 00:00:00 0 8 I 1/SAMGR: Init service 0x4ae760 <time: 0ms> success!
00 00:00:00 0 8 I 1/SAMGR: Initialized all core system services!
life value:1868
life value:1866
life value:128
life value:128
life value:128
```

图 2-19 生命探测仪的运行效果

2.2.7 案例 8：土壤湿度监测仪

本案例讲解土壤湿度监测仪，实现土壤湿度监测的功能，其中核心模块为土壤湿度传感器模块，如图 2-20 所示。

图 2-20 中左面为土壤湿度检测头，右面为土壤湿度信号处理模块，它们之间通过红、棕两根母对母的杜邦线连接。信号处理模块下面的 4 个引脚分别为 AO、DO、GND、VCC，除 DO 外分别与主控板的 GPIO09(ADC4)、GND、+5V 相连。

土壤湿度传感器模块可以宽范围控制土壤的湿度，通过电位器调节控制的相应阈值，当湿度低于设定值时，DO 输出高电平，当高于设定值时，DO 输出低电平，也可以通过 AO 输出模拟信号值。核心芯片为 LM393 芯片，工作电压 3.3～5V。

开发步骤如下：

（1）在 chapter_02 中创建工程目录 case8-soil_moisture_detector。

（2）在 case8-soil_moisture_detector 中创建源码文件 soil_moisture_detector_demo.c。

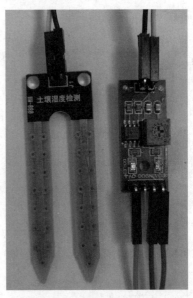

图 2-20 土壤湿度传感器模块

（3）功能实现，代码如下：

```
//applications/sample/wifi-iot/app/ohos_50/chapter_02/case8-soil_moisture_detector/soil_moisture_detector_demo.c
```

```c
#include <stdio.h>
#include <unistd.h>
#include <ohos_init.h>
#include <wifiiot_adc.h>
//将使用的 ADC CHANNEL 序号声明为 4
#define ADC_LIGHT WIFI_IOT_ADC_CHANNEL_4
//用来保存读取到的模拟数据
unsigned short data = 0;
void entry(void)
{
    sleep(1);
    while (1)
    {
        AdcRead(ADC_LIGHT, &data, WIFI_IOT_ADC_EQU_MODEL_4, WIFI_IOT_ADC_CUR_BAIS_DEFAULT, 0);
        printf("soil moisture value: %d\n", data);
        usleep(300000);
    }
}
APP_FEATURE_INIT(entry);
```

(4) 在 case8-soil_moisture_detector 中创建模块，构建脚本 BUILD.gn 并初始化模块，代码如下：

```
#//applications/sample/wifi-iot/app/ohos_50/chapter_02/case8-soil_moisture_detector/BUILD.gn
static_library("ch_02_soil_moisture_detector") {
    sources = [
        "soil_moisture_detector_demo.c",
    ]
    include_dirs = [
        "//base/iot_hardware/interfaces/kits/wifiiot_lite",
    ]
}
```

(5) 将模块 ch_02_soil_moisture_detector 配置到应用子系统，代码如下：

```
#//applications/sample/wifi-iot/app/BUILD.gn
import("//build/lite/config/component/lite_component.gni")
lite_component("app") {
    features = [
        "ohos_50/chapter_02/case8-soil_moisture_detector:ch_02_soil_moisture_detector",
    ]
}
```

(6) 测试：编译应用模块，将固件烧写到开发板，运行 IPOP 终端工具，以便与开发板相连，复位开发板，向土壤滴水。观察 IPOP 终端工具的打印信息，如图 2-21 所示。

```
ready to OS start
sdk ver:Hi3861V100R001C00SPC025 2020-09-03 18:10:00
FileSystem mount ok.
wifi init success!

00 00:00:00 0 132 D 0/HIVIEW: hilog init success.
00 00:00:00 0 132 D 0/HIVIEW: log limit init success.
00 00:00:00 0 132 I 1/SAMGR: Bootstrap core services(count:3).
00 00:00:00 0 132 I 1/SAMGR: Init service:0x4ae61c TaskPool:0xfa1e4
00 00:00:00 0 132 I 1/SAMGR: Init service:0x4ae640 TaskPool:0xfa854
00 00:00:00 0 132 I 1/SAMGR: Init service:0x4ae768 TaskPool:0xfaa14
00 00:00:00 0 164 I 1/SAMGR: Init service 0x4ae640 <time: 0ms> success
00 00:00:00 0 64 I 1/SAMGR: Init service 0x4ae61c <time: 0ms> success
00 00:00:00 0 8 D 0/HIVIEW: hiview init success.
00 00:00:00 0 8 I 1/SAMGR: Init service 0x4ae768 <time: 0ms> success!
00 00:00:00 0 8 I 1/SAMGR: Initialized all core system services!
soil moisture value:600
soil moisture value:592
soil moisture value:593
```

图 2-21　土壤湿度监测仪的运行效果

2.2.8　案例 9：电压调节器

本案例讲解电压调节器，实现可变电压输出功能，其中核心模块为电位器模块，如图 2-22 所示。

右面为电位器，左面的 3 个引脚分别为 GND、VCC、OUT，分别与主控板的 GND、+5V、GPIO011（ADC5）相连。

电位器是具有 3 个引出端且阻值可按某种变化规律调节的电阻元件。电位器通常由电阻体和可移动的电刷组成。当电刷沿电阻体移动时，在输出端即可获得与位移量成一定关系的电阻值或电压。

图 2-22　电位器模块

开发步骤如下：

(1) 在 chapter_02 中创建工程目录 case9-voltage_regulator。

(2) 在 case9-voltage_regulator 中创建源码文件 voltage_regulator_demo.c。

(3) 功能实现，代码如下：

```
//applications/sample/wifi-iot/app/ohos_50/chapter_02/case9-voltage_regulator/voltage_regulator_demo.c
#include <stdio.h>
#include <unistd.h>
#include <ohos_init.h>
```

```c
#include <wifiiot_adc.h>
//将使用的 ADC CHANNEL 序号声明为 5
#define ADC_LIGHT WIFI_IOT_ADC_CHANNEL_5
//用来保存读取到的模拟数据
unsigned short data = 0;
void entry(void)
{
    sleep(1);
    while (1)
    {
        AdcRead(ADC_LIGHT, &data, WIFI_IOT_ADC_EQU_MODEL_4, WIFI_IOT_ADC_CUR_BAIS_DEFAULT, 0);
        printf("voltage value:% d\n", data);
        usleep(300000);
    }
}
APP_FEATURE_INIT(entry);
```

（4）在 case9-voltage_regulator 中创建模块，构建脚本 BUILD.gn 并初始化模块，代码如下：

```
#//applications/sample/wifi-iot/app/ohos_50/chapter_02/case9-voltage_regulator/BUILD.gn
static_library("ch_02_voltage_regulator") {
    sources = [
        "voltage_regulator_demo.c",
    ]
    include_dirs = [
        "//base/iot_hardware/interfaces/kits/wifiiot_lite",
    ]
}
```

（5）将模块 ch_02_voltage_regulator 配置到应用子系统，代码如下：

```
#//applications/sample/wifi-iot/app/BUILD.gn
import("//build/lite/config/component/lite_component.gni")
lite_component("app") {
    features = [
        "ohos_50/chapter_02/case9-voltage_regulator:ch_02_voltage_regulator",
    ]
}
```

（6）测试：编译应用模块，将固件烧写到开发板，运行 IPOP 终端工具，以便与开发板相连，复位开发板，转动电位器的转轴。观察 IPOP 终端工具的打印信息，如图 2-23 所示。

```
ready to OS start
sdk ver:Hi3861V100R001C00SPC025 2020-09-03 18:10:00
formatting spiffs...
FileSystem mount ok.
wifi init success!
00 00:00:00 0 132 D 0/HIVIEW: hilog init success.
00 00:00:00 0 132 D 0/HIVIEW: log limit init success.
00 00:00:00 0 132 I 1/SAMGR: Bootstrap core services(count:3).
00 00:00:00 0 132 I 1/SAMGR: Init service:0x4ae61c TaskPool:0xfa1e4
00 00:00:00 0 132 I 1/SAMGR: Init service:0x4ae640 TaskPool:0xfa854
00 00:00:00 0 132 I 1/SAMGR: Init service:0x4ae764 TaskPool:0xfaa14
00 00:00:00 0 164 I 1/SAMGR: Init service 0x4ae640 <time: 0ms> succes
00 00:00:00 0 64 I 1/SAMGR: Init service 0x4ae61c <time: 0ms> success
00 00:00:00 0 8 D 0/HIVIEW: hiview init success.
00 00:00:00 0 8 I 1/SAMGR: Init service 0x4ae764 <time: 0ms> success!
00 00:00:00 0 8 I 1/SAMGR: Initialized all core system services!
voltage value:128
voltage value:132
voltage value:152
voltage value:192
voltage value:341
voltage value:628
voltage value:1764
```

图 2-23 噪声检测仪电压调节器的运行效果

2.3 GPIO

通用型输入/输出(General-Purpose Input/Output,GPIO)包含通用输出(GPO)和通用输入(GPI)。Hi3861 芯片有 15 个支持 GPIO 的引脚,这些引脚同时还支持其他的硬件接口,如 I^2C、ADC、I^2S、PWM、UART 等。

Hi3861 芯片的 GPIO 设备接口的开发流程如下:

(1) 分析电路,确定 GPIO 引脚索引及 I/O 方向。

(2) 调用函数 GpioInit 初始化 GPIO 引脚。

(3) 调用函数 IoSetFunc 将引脚的功能设置为 GPIO。

(4) 调用函数 GpioSetDir 将引脚设置为输入或输出。

(5) 如果是输入,则调用函数 GpioGetInputVal 获取引脚的数字信息,结果为 0(表示低电平,0V)或 1(表示高电平,在 Hi3861 开发板上高电平为 3.3V)。

(6) 如果是输出,则调用函数 GpioSetOutputVal 向引脚输出 0 或 1。

2.3.1 案例 10:工作指示灯

本案例为工作指示灯,实现主控板上的 D2 发光二极管亮灭的控制功能。D2 如图 2-24 所示。

D2 发光二极管与 Hi3861 的 GPIO09 引脚相连,电路图如图 2-25 所示。

左面接入 3.3V 高电平,然后连接一个 470Ω 的电阻,再与 D2 发光二极管相连,由图可知,需要连接 J8 跳线帽才能使 D2 发光二极管与 Hi3861 芯片的 GPIO9 引脚相连通。

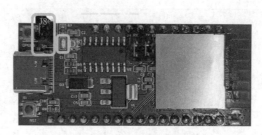

图 2-24　D2 LED 工作指示灯

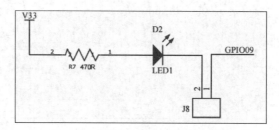

图 2-25　D2 LED 工作指示灯电路图

发光二极管具有单向导通性。如果想点亮发光二极管,则电流只能从左边的正极流入,从右边的负极流出,所以只有 GPIO09 引脚输出低电平时电流才能通过 D2 发光二极管,D2 发光二极管被点亮。

本案例的实现原理是,根据 GPIO09 引脚输出的高低电平控制 D2 发光二极管的灭亮,亮表示工作,灭代表空闲。再通过随机数来模拟空闲时长。

开发步骤如下:

(1) 在 chapter_02 中创建工程目录 case10-indicator。

(2) 在 case10-indicator 中创建源码文件 indicator_demo.c。

(3) 功能实现,代码如下:

```c
//applications/sample/wifi-iot/app/ohos_50/chapter_02/case10-indicator/indicator_demo.c
#include <stdio.h>
#include <unistd.h>
#include <stdlib.h>
#include <ohos_init.h>
#include "wifiiot_gpio.h"
#include "wifiiot_gpio_ex.h"
/*
 *实现思路:
    1.初始化
        (1)分析电路,确定 GPIO 引脚索引为 WIFI_IOT_IO_NAME_GPIO_9
        (2)调用函数 GpioInit 初始化 GPIO 引脚
```

(3)调用函数 IoSetFunc 将引脚的功能设置为 GPIO
　　　(4)调用函数 GpioSetDir 将引脚设置为输出
　2. 功能实现
　　　(1)循环调用函数 GpioSetOutputVal 输出低或高电平,实现 LED 亮灭功能
 *
 */
```c
//初始化引脚 WIFI_IOT_IO_NAME_GPIO_9
#define PIN WIFI_IOT_IO_NAME_GPIO_9
void init(void)
{
    GpioInit();            //初始化 GPIO 设备,使用 GPIO 引脚前调用
    //将 GPIO_9 的功能设置为 IO
    IoSetFunc(PIN, WIFI_IOT_IO_FUNC_GPIO_9_GPIO);
    //将 GPIO_9 的功能设置为输出
    GpioSetDir(PIN, WIFI_IOT_GPIO_DIR_OUT);
}
//工作指示灯功能实现
void entry(void)
{
    init();
    long delay;            //保存随机数
    while (1)
    {
        //将 GPIO_9 的输出值设置为低电平,值为 0
        GpioSetOutputVal(PIN, WIFI_IOT_GPIO_VALUE0);
        usleep(200000);    //等待 0.2s
        //将 GPIO_9 的输出值设置为高电平,LED 熄灭
        GpioSetOutputVal(PIN, WIFI_IOT_GPIO_VALUE1);
        //获取 10 以内的随机数
        delay = rand() % 10;
        //休眠 delay * 100000 微秒
        usleep(delay * 100000);
    }
}
//将模块入口函数初始化为 entry
APP_FEATURE_INIT(entry);
```

(4) 在 case10-indicator 中创建模块,构建脚本 BUILD.gn 并初始化模块,代码如下:

```
#//applications/sample/wifi-iot/app/ohos_50/chapter_02/case10-indicator/BUILD.gn
static_library("ch_02_indicator") {
    sources = [
        "indicator_demo.c",
    ]
    include_dirs = [
```

```
        "//base/iot_hardware/interfaces/kits/wifiiot_lite",
    ]
}
```

（5）将模块 ch_02_indicator 配置到应用子系统，代码如下：

```
#//applications/sample/wifi-iot/app/BUILD.gn
import("//build/lite/config/component/lite_component.gni")
lite_component("app") {
    features = [
        "ohos_50/chapter_02/case10-indicator:ch_02_indicator",
    ]
}
```

（6）测试：编译应用模块，将固件烧写到开发板，复位开发板，D2 发光二极管会无规律地闪烁，如图 2-26 所示。

图 2-26　D2 工作指示灯的运行效果

2.3.2　案例 11：智能开关

本案例为智能开关，实现控制继电器开关控制功能，核心模块为继电器模块 JQC-3FF-S-Z，如图 2-27 所示。

图 2-27　JQC-3FF-S-Z 继电器

左面引脚 DC＋与主控板 5V 电源相连,引脚 DC－与主控板的 GND 相连,引脚 IN 与主控板的 GPIO10 引脚相连。

右面引脚 COM 为控制线的输入端,NO(Normal Open)是常开输出端,继电器线圈未通电时断开；NC(Normal Close)是常闭输出端,继电器线圈未通电时导通。

继电器(英文名称:relay)是一种电控制器件,当输入量(激励量)的变化达到规定要求时,在电气输出电路中使被控量发生预定的阶跃变化的一种电器。它具有控制系统(又称输入回路)和被控制系统(又称输出回路)之间的互动关系。通常应用于自动化的控制电路中,它实际上是用小电流去控制大电流运作的一种"自动开关"。故在电路中起着自动调节、安全保护、转换电路等作用。

案例中使用的继电器的型号为 JQC-3FF-S-Z,兼容 3.3V 信号驱动,可以通过设置跳线使用高电平触发继电器。

开发步骤如下:

(1) 在 chapter_02 中创建工程目录 case11-relay。

(2) 在 case11-relay 中创建源码文件 relay_demo.c。

(3) 功能实现,代码如下:

```c
//applications/sample/wifi-iot/app/ohos_50/chapter_02/case11-relay/relay_demo.c
#include <stdio.h>
#include <unistd.h>
#include <stdlib.h>
#include "ohos_init.h"
#include "wifiiot_gpio.h"
#include "wifiiot_gpio_ex.h"
#define Pin WIFI_IOT_IO_NAME_GPIO_10
void entry(void)
{
    GpioInit();                    //初始化 GPIO 设备,使用 GPIO 引脚前调用
    //将 GPIO_10 功能设置为 IO
    IoSetFunc(Pin, WIFI_IOT_IO_FUNC_GPIO_10_GPIO);
    //将 GPIO_10 的功能设置为输出
    GpioSetDir(Pin, WIFI_IOT_GPIO_DIR_OUT);
    while (1)
    {
        //高电平触发继电器
        GpioSetOutputVal(Pin, WIFI_IOT_GPIO_VALUE1);
        sleep(rand() % 3 + 1);   //随机等待[0,3]秒
        //低电平断开继电器
        GpioSetOutputVal(Pin, WIFI_IOT_GPIO_VALUE0);
        sleep(rand() % 3 + 1);   //随机等待[0,3]秒
    }
}
APP_FEATURE_INIT(entry);
```

（4）在case11-relay中创建模块，构建脚本BUILD.gn并初始化模块，代码如下：

```
#//applications/sample/wifi-iot/app/ohos_50/chapter_02/case11-relay/BUILD.gn
static_library("ch_02_relay") {
    sources = [
        "relay_demo.c",
    ]
    include_dirs = [
        "//base/iot_hardware/interfaces/kits/wifiiot_lite",
    ]
}
```

（5）将模块ch_02_relay配置到应用子系统，代码如下：

```
//applications/sample/wifi-iot/app/BUILD.gn
import("//build/lite/config/component/lite_component.gni")
lite_component("app") {
    features = [
        "ohos_50/chapter_02/case11-relay:ch_02_relay",
    ]
}
```

（6）测试：将继电器与主控板相连，编译应用模块，将固件烧写到开发板，复位开发板，继电器会无规律地连接与断开并伴随着"啪"声，继电器状态指示灯也会相应地亮灭，如图2-28所示。

图2-28 继电器的运行效果

2.3.3 案例12：SOS摩斯密码发射器

本案例是SOS摩斯密码发射器，通过控制激光发射器实现SOS摩斯密码发射功能，其中核心模块为激光发射器模块，如图2-29所示。

上面为激光发射头，下面的3个引脚从左往右依次为信号、5V VCC、GND。分别与主控板的 GPIO11、+5V、GND 相连，其中 GPIO11 为激光发射控制信号线，通过 GPIO11 输出高电平触发。

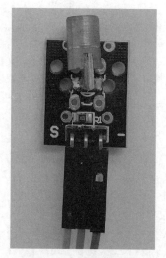

图2-29 激光发射器

开发步骤如下：

(1) 在 chapter_02 中创建工程目录 case12-laser。
(2) 在 case12-laser 中创建源码文件 laser_demo.c。
(3) 功能实现，代码如下：

```c
//applications/sample/wifi-iot/app/ohos_50/chapter_02/case12-laser/laser_demo.c
#include <stdio.h>
#include <unistd.h>
#include <stdlib.h>
#include "ohos_init.h"
#include "wifiiot_gpio.h"
#include "wifiiot_gpio_ex.h"

//激光信号控制引脚
#define Pin WIFI_IOT_IO_NAME_GPIO_11
//摩斯密码 "嘀"周期，单位为微秒
#define T 300000
/**
 * 本案例通过激光发射器打出摩斯密码 SOS
 * "SOS"对应的摩斯密码为... --- ...
 * . 发嘀,时长为1T
 * - 发嗒,时长为3T
 * "SOS"摩斯密码转换为数组为{{1,1,1},{3,3,3},{1,1,1}}
 */
void entry(void)
{
    //SOS 摩斯密码数组
    int sos_array[3][3] = {
        {1, 1, 1}, //'S'摩斯密码(...)
        {3, 3, 3}, //'O'摩斯密码(---)
        {1, 1, 1}  //'S'摩斯密码(...)
    };
    GpioInit();
```

```c
        IoSetFunc(Pin, WIFI_IOT_IO_FUNC_GPIO_11_GPIO);
        GpioSetDir(Pin, WIFI_IOT_GPIO_DIR_OUT);
        while (1)
        {
            //循环输出 SOS 摩斯密码信号
            for (int i = 0; i < 3; i++)
            {
                for (int j = 0; j < 3; j++)
                {
                    GpioSetOutputVal(Pin, WIFI_IOT_GPIO_VALUE1);
                    usleep(sos_array[i][j] * T);
                    GpioSetOutputVal(Pin, WIFI_IOT_GPIO_VALUE0);
                    usleep(T);
                }
                usleep(T);
            }
            usleep(T * 3);
        }
}
APP_FEATURE_INIT(entry);
```

(4) 在 case12-laser 中创建模块,构建脚本 BUILD.gn 并初始化模块,代码如下:

```
#//applications/sample/wifi-iot/app/ohos_50/chapter_02/case12-laser/BUILD.gn
static_library("ch_02_laser") {
    sources = [
        "laser_demo.c",
    ]
    include_dirs = [
        "//base/iot_hardware/interfaces/kits/wifiiot_lite",
    ]
}
```

(5) 将模块 ch_02_laser 配置到应用子系统,代码如下:

```
#//applications/sample/wifi-iot/app/BUILD.gn
import("//build/lite/config/component/lite_component.gni")
lite_component("app") {
    features = [
        "ohos_50/chapter_02/case12-laser:ch_02_laser",
    ]
}
```

(6) 测试:将激光发射器与主控板相连,编译应用模块,将固件烧写到开发板,复位开发板,激光发射器会循环发出与 SOS 摩斯密码相应的激光信号,如图 2-30 所示。

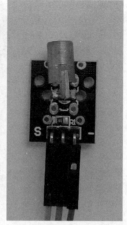

图 2-30　SOS 摩斯密码发射器的运行效果

2.3.4　案例 13：倾斜检测仪

本案例是倾斜检测仪，实现倾斜探测功能，核心模块为倾斜开关模块，如图 2-31 所示。

上面为倾斜检测头，下面的 3 个引脚从左往右依次为信号、GND、5V VCC。分别与主控板的 GPIO12、GND、+5V 相连，其中 GPIO12 为倾斜检测信号线，当 GPIO12 输入 0 时表示倾斜，当输入 1 时表示水平。

在倾斜开关中球以不同的倾斜角度移动以触发电路。倾斜开关模块使用双向传导的球形倾斜开关。当它向任意一侧倾斜时，只要倾斜度和力满足条件，开关就会通电，从而输出低电平信号。

开发步骤如下：

（1）在 chapter_02 中创建工程目录 case13-tilt_switch。

（2）在 case13-tilt_switch 中创建源码文件 tilt_switch_demo.c。

（3）功能实现，代码如下：

图 2-31　倾斜开关

```
//applications/sample/wifi-iot/app/ohos_50/chapter_02/case13-tilt_switch/tilt_switch_demo.c
#include <stdio.h>
#include <unistd.h>
#include <ohos_init.h>
#include "wifiiot_gpio.h"
```

```c
#include "wifiiot_gpio_ex.h"
#define Pin WIFI_IOT_IO_NAME_GPIO_12
void entry(void)
{
    GpioInit();
    IoSetFunc(Pin, WIFI_IOT_IO_FUNC_GPIO_12_GPIO);
    GpioSetDir(Pin, WIFI_IOT_GPIO_DIR_IN);

    //声明 wifiiotGpioValue 变量 value,用来保存 GPIO12 引脚的高低电平数值
    wifiiotGpioValue value;
    sleep(1);
    printf("\nfilt switch stat:1");
    while (1)
    {
        usleep(50000);
        //读取 GPIO_12 的值
        GpioGetInputVal(Pin, &value);
        printf("\b%d",value);
    }
}
APP_FEATURE_INIT(entry);
```

（4）在 case13-tilt_switch 中创建模块,构建脚本 BUILD.gn 并初始化模块,代码如下：

```
#//applications/sample/wifi-iot/app/ohos_50/chapter_02/case13-tilt_switch/BUILD.gn
static_library("ch_02_tilt") {
    sources = [
        "tilt_switch_demo.c",
    ]
    include_dirs = [
        "//base/iot_hardware/interfaces/kits/wifiiot_lite",
    ]
}
```

（5）将模块 ch_02_tilt 配置到应用子系统,代码如下：

```
#//applications/sample/wifi-iot/app/BUILD.gn
import("//build/lite/config/component/lite_component.gni")
lite_component("app") {
    features = [
        "ohos_50/chapter_02/case13-tilt_switch:ch_02_tilt",
    ]
}
```

(6) 测试：将倾斜开关与主控板相连，编译应用模块，将固件烧写到开发板，运行 IPOP 终端工具，以便与开发板相连，复位开发板，使倾斜开关倾斜，观察 IPOP 终端工具中的打印信息，如图 2-32 所示。

```
? ready to OS start
sdk ver:Hi3861V100R001C00SPC025 2020-09-03 18:10:00
FileSystem mount ok.
wifi init success!

00 00:00:00 0 132 D 0/HIVIEW: hilog init success.
00 00:00:00 0 132 D 0/HIVIEW: log limit init success.
00 00:00:00 0 132 I 1/SAMGR: Bootstrap core services(count:3).
00 00:00:00 0 132 I 1/SAMGR: Init service:0x4ae71c TaskPool:0xfa1e4
00 00:00:00 0 132 I 1/SAMGR: Init service:0x4ae740 TaskPool:0xfa854
00 00:00:00 0 132 I 1/SAMGR: Init service:0x4ae868 TaskPool:0xfaa14
00 00:00:00 0 164 I 1/SAMGR: Init service 0x4ae740 <time: 0ms> succes
00 00:00:00 0 64 I 1/SAMGR: Init service 0x4ae71c <time: 0ms> success
00 00:00:00 0 8 D 0/HIVIEW: hiview init success.
00 00:00:00 0 8 I 1/SAMGR: Init service 0x4ae868 <time: 0ms> success!
00 00:00:00 0 8 I 1/SAMGR: Initialized all core system services!

filt switch stat:0_
```

图 2-32 倾斜检测仪的运行效果

2.3.5 案例 14：地震监测仪

本案例为地震监测仪，实现震动监测的功能，其中核心模块为振动开关模块，如图 2-33 所示。

上面为振动检测头，下面的 3 个引脚从左往右依次为信号、GND、VCC。分别与主控板的 GPIO13、GND、+5V 相连，其中 GPIO13 为振动检测信号线，当 GPIO13 输入 1 时表示振动，当输入 0 时表示静止。

在振动开关模块中，导电的振动弹簧和触发销被精确地放置在开关体中并且通过黏合剂结合到固化位置。通常，弹簧和触发销不接触。一旦摇动，弹簧就会摇动并与触发器引脚接触以传导并产生触发信号。

图 2-33 振动开关

开发步骤如下：

(1) 在 chapter_02 中创建工程目录 case14-vibration_switch。

(2) 在 case14-vibration_switch 中创建源码文件 vibration_switch_demo.c。

(3) 功能实现，代码如下：

//applications/sample/wifi-iot/app/ohos_50/chapter_02/case14-vibration_switch/vibration_switch_demo.c

```c
#include <stdio.h>
#include <unistd.h>
#include <ohos_init.h>
#include "wifiiot_gpio.h"
#include "wifiiot_gpio_ex.h"
#define Pin WIFI_IOT_IO_NAME_GPIO_13
void entry(void)
{
    GpioInit();
    IoSetFunc(Pin, WIFI_IOT_IO_FUNC_GPIO_13_GPIO);
    GpioSetDir(Pin, WIFI_IOT_GPIO_DIR_IN);
    //声明变量value,用来保存GPIO13引脚的高低电平数值
    wifiiotGpioValue value;
    sleep(1);
    //静止时状态为0(低电平),振动时状态为1(高电平)
    printf("\nvibration switch stat:0");
    while (1)
    {
        usleep(50000);
        //读取GPIO_13的值
        GpioGetInputVal(Pin, &value);
        printf("\b%d",value);
    }
}
APP_FEATURE_INIT(entry);
```

（4）在case14-vibration_switch中创建模块,构建脚本BUILD.gn并初始化模块,代码如下：

```
#//applications/sample/wifi-iot/app/ohos_50/chapter_02/case14-vibration_switch/BUILD.gn
static_library("ch_02_vibration") {
    sources = [
        "vibration_switch_demo.c",
    ]
    include_dirs = [
        "//base/iot_hardware/interfaces/kits/wifiiot_lite",
    ]
}
```

（5）将模块ch_02_vibration配置到应用子系统,代码如下：

```
#//applications/sample/wifi-iot/app/BUILD.gn
import("//build/lite/config/component/lite_component.gni")
lite_component("app") {
```

```
        features = [
            "ohos_50/chapter_02/case14-vibration_switch:ch_02_vibration",
        ]
}
```

（6）测试：将振动开关与主控板相连，编译应用模块，将固件烧写到开发板，运行 IPOP 终端工具，以便与开发板相连，复位开发板，使振动开关振动，观察 IPOP 终端工具中的打印信息，如图 2-34 所示。

图 2-34　地震检测仪的运行效果

2.3.6　案例 15：机械手臂

本案例是机械手臂，实现伺服电机的旋转功能，其中核心模块为伺服舵机，如图 2-35 所示。

上面为振动检测头，左面的 3 个引脚从前往后依次为 GND、VCC、信号，分别与主控板的 GND、+5V、GPIO14 相连，其中 GPIO14 为舵机控制信号线。

图 2-35　伺服舵机

舵机的控制一般需要一个 20ms 左右的时基脉冲，该脉冲的高电平部分一般为 0.5~2.5ms 的角度控制脉冲部分，总间隔为 2ms。以 180°角度伺服舵机为例，对应的控制关系如表 2-1 所示。

表 2-1　输入高电平时长与舵机转动角度的关系

输入高电平时长/ms	舵机转动角度/(°)
0.5	0
1.0	45
1.5	90
2.0	135
2.5	180

开发步骤如下：
(1) 在 chapter_02 中创建工程目录 case15-servo。
(2) 在 case15-servo 中创建源码文件 servo_demo.c。
(3) 功能实现，代码如下：

```c
//applications/sample/wifi-iot/app/ohos_50/chapter_02/case15-servo/servo_demo.c
#include <stdio.h>
#include <unistd.h>
#include <ohos_init.h>
#include "wifiiot_gpio.h"
#include "wifiiot_gpio_ex.h"
#define Pin WIFI_IOT_IO_NAME_GPIO_14
void entry(void)
{
    GpioInit();
    IoSetFunc(Pin, WIFI_IOT_IO_FUNC_GPIO_14_GPIO);
    GpioSetDir(Pin, WIFI_IOT_GPIO_DIR_OUT);
    while(1){
        for(int i = 1000; i <= 2000; i += 2){
            GpioSetOutputVal(Pin, WIFI_IOT_GPIO_VALUE1);
            usleep(i);
            GpioSetOutputVal(Pin, WIFI_IOT_GPIO_VALUE0);
            usleep(10000);
        }
        for(int i = 2000; i >= 1000; i -= 2){
            GpioSetOutputVal(Pin, WIFI_IOT_GPIO_VALUE1);
            usleep(i);
            GpioSetOutputVal(Pin, WIFI_IOT_GPIO_VALUE0);
            usleep(10000);
        }
    }
}
APP_FEATURE_INIT(entry);
```

(4) 在 case15-servo 中创建模块，构建脚本 BUILD.gn 并初始化模块，代码如下：

```
#//applications/sample/wifi-iot/app/ohos_50/chapter_02/case15-servo/BUILD.gn
static_library("ch_02_servo") {
    sources = [
        "servo_demo.c",
    ]
    include_dirs = [
        "//base/iot_hardware/interfaces/kits/wifiiot_lite",
    ]
}
```

(5)将模块 ch_02_servo 配置到应用子系统,代码如下:

```
#//applications/sample/wifi-iot/app/BUILD.gn
import("//build/lite/config/component/lite_component.gni")
lite_component("app") {
    features = [
        "ohos_50/chapter_02/case15-servo:ch_02_servo",
    ]
}
```

(6)测试:将舵机与主控板相连,编译应用模块,将固件烧写到开发板,复位开发板,观察舵机转动的角度,如图 2-36 所示。

图 2-36　伺服舵机运行效果

2.3.7　案例 16:缝隙探测器

本案例是缝隙探测器,实现缝隙探测的功能,其中核心模块为寻迹模块,如图 2-37 所示。

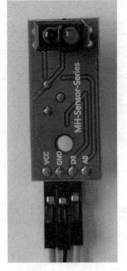

图 2-37　寻迹模块

正面为信号处理模块,背面为寻迹探头。下面 4 个引脚从左往右依次为 AO、DO、GND、VCC。除 DO 外分别与主控板的 GPIO00、GND、+5V 相连,其中 AO 不与主控板相连,GPIO00 为缝隙检测信号线,当 GPIO00 输入 1 时表示有缝隙,当输入 0 时表示无缝隙。

本模块的红外发射二极管不断地发射红外线,当发射出的红外线没有被反射回来或被反射回来但强度不够大时,光敏三极管一直处于关断状态,此时模块的输出端为低电平,指示二极管一直处于熄灭状态;当被检测物体出现在检测范围内时,红外线被反射回来且强度足够大,光敏三极管饱和,此时模块的输出端为高电平,指示二极管被点亮。

开发步骤如下:

(1) 在 chapter_02 中创建工程目录 case16-slot_detector。

(2) 在 case16-slot_detector 中创建源码文件 slot_detector_demo.c。

(3) 功能实现,代码如下:

```c
//applications/sample/wifi-iot/app/ohos_50/chapter_02/case16-slot_detector/slot_detector_demo.c
#include <stdio.h>
#include <unistd.h>
#include <ohos_init.h>
#include "wifiiot_gpio.h"
#include "wifiiot_gpio_ex.h"
#define Pin WIFI_IOT_IO_NAME_GPIO_0
void entry(void)
{
    GpioInit();
    IoSetFunc(Pin, WIFI_IOT_IO_FUNC_GPIO_0_GPIO);
    GpioSetDir(Pin, WIFI_IOT_GPIO_DIR_IN);
    wifiiotGpioValue value;
    sleep(1);
    printf("\nslot value:1");
    while(1){
        usleep(50000);
        GpioGetInputVal(Pin, &value);
        printf("\b%d",value);
    }
}
APP_FEATURE_INIT(entry);
```

(4) 在 case16-slot_detector 中创建模块,构建脚本 BUILD.gn 并初始化模块,代码如下:

```
#//applications/sample/wifi-iot/app/ohos_50/chapter_02/case16-slot_detector/BUILD.gn
static_library("ch_02_slot") {
    sources = [
        "slot_detector_demo.c",
    ]
```

```
    include_dirs = [
        "//base/iot_hardware/interfaces/kits/wifiiot_lite",
    ]
}
```

(5) 将模块 ch_02_slot 配置到应用子系统,代码如下:

```
#//applications/sample/wifi-iot/app/BUILD.gn
import("//build/lite/config/component/lite_component.gni")
lite_component("app") {
    features = [
        "ohos_50/chapter_02/case16-slot_detector:ch_02_slot",
    ]
}
```

(6) 测试:将寻迹模块与主控板相连,编译应用模块,将固件烧写到开发板,运行 IPOP 终端工具,以便与开发板相连,复位开发板,使寻迹探头下方有 5cm 以上空间,观察 IPOP 终端工具中的打印信息,如图 2-38 所示。

图 2-38　缝隙探测器的运行效果

2.3.8　案例 17:触摸感应器

本案例讲解触摸感应器,实现人体触摸感应功能。核心模块为触摸感应器,如图 2-39 所示。

触摸感应器默认输出低电平,当手指触摸时输出高电平,其作用类似于轻触按键。共 3 个引脚(GND、VCC、SIG),GND 为地,VCC 为供电电源,SIG 为数字信号输出引脚。

触摸感应器实现触摸状态读取功能。

图 2-39　触摸传感器

开发步骤如下：

（1）在 chapter_02 中创建工程目录 case17-touch_sensor。

（2）在 case17-touch_sensor 中创建源码文件 touch_sensor_demo.c。

（3）功能实现，代码如下：

```c
//applications/sample/wifi-iot/app/ohos_50/chapter_02/case17-touch_sensor/touch_sensor_demo.c
#include <stdio.h>
#include <unistd.h>
#include <ohos_init.h>
#include "wifiiot_gpio.h"
#include "wifiiot_gpio_ex.h"
#define Pin WIFI_IOT_IO_NAME_GPIO_1
void entry(void)
{
    GpioInit();
    IoSetFunc(Pin, WIFI_IOT_IO_FUNC_GPIO_1_GPIO);
    GpioSetDir(Pin, WIFI_IOT_GPIO_DIR_IN);
    wifiiotGpioValue value;
    sleep(1);
    printf("\ntouch value:1");
    while(1){
        usleep(500000);
        GpioGetInputVal(Pin, &value);
        printf("\b%d",value);
    }
}
APP_FEATURE_INIT(entry);
```

（4）在 case17-touch_sensor 中创建模块，构建脚本 BUILD.gn 并初始化模块，代码如下：

```
#//applications/sample/wifi-iot/app/ohos_50/chapter_02/case17-touch_sensor/BUILD.gn
static_library("ch_02_touch") {
    sources = [
        "touch_sensor_demo.c",
    ]
    include_dirs = [
        "//base/iot_hardware/interfaces/kits/wifiiot_lite",
    ]
}
```

（5）将模块 ch_02_touch 配置到应用子系统，代码如下：

```
#//applications/sample/wifi-iot/app/BUILD.gn
import("//build/lite/config/component/lite_component.gni")
lite_component("app") {
```

```
    features = [
        "ohos_50/chapter_02/case17-touch_sensor:ch_02_touch",
    ]
}
```

(6)测试:将触摸传感器与主控板引脚相连,编译应用模块,将固件烧写到开发板,运行 IPOP 终端工具,以便与开发板相连,复位开发板,将手指放在触摸感应器上,观察 IPOP 终端工具中的打印信息,如图 2-40 所示。

```
ready to OS start
sdk ver:Hi3861V100R001C00SPC025 2020-09-03 18:10:00
FileSystem mount ok.
wifi init success!
00 00:00:00 0 132 D 0/HIVIEW: hilog init success.
00 00:00:00 0 132 D 0/HIVIEW: log limit init success.
00 00:00:00 0 132 I 1/SAMGR: Bootstrap core services(count:3).
00 00:00:00 0 132 I 1/SAMGR: Init service:0x4ae71c TaskPool:0xfa1e4
00 00:00:00 0 132 I 1/SAMGR: Init service:0x4ae740 TaskPool:0xfa854
00 00:00:00 0 132 I 1/SAMGR: Init service:0x4ae864 TaskPool:0xfaa14
00 00:00:00 0 164 I 1/SAMGR: Init service 0x4ae740 <time: 0ms> succes
00 00:00:00 0 64 I 1/SAMGR: Init service 0x4ae71c <time: 0ms> success
00 00:00:00 0 8 D 0/HIVIEW: hiview init success.
00 00:00:00 0 8 I 1/SAMGR: Init service 0x4ae864 <time: 0ms> success!
00 00:00:00 0 8 I 1/SAMGR: Initialized all core system services!
touch value:1_
```

图 2-40 触摸感应器的运行效果

2.3.9 案例 18:火焰探测器

本案例为火焰探测器,实现火焰探测功能,其中核心模块为火焰传感器,如图 2-41 所示。

上面为火焰探测头,下面 4 个引脚从左往右依次为 AO、DO、GND、VCC。除 AO 外分别与主控板的 GPIO02、+5V、GND 相连,其中 AO 不与主控板相连,GPIO02 读取的值即为火焰状态值,当 GPIO02 输入 0 时表示有火焰,当输入 1 时表示无火焰。

火焰探测器通过读取信号引脚的数字信号来输出火焰状态。开发步骤如下:

(1)在 chapter_02 中创建工程目录 case18-flame_detector。

(2)在 case18-flame_detector 中创建源码文件 flame_detector_demo.c。

(3)功能实现,代码如下:

图 2-41 火焰传感器

//applications/sample/wifi-iot/app/ohos_50/chapter_02/case18-flame_detector/flame_detector_demo.c

```c
#include <stdio.h>
#include <unistd.h>
#include <ohos_init.h>
#include "wifiiot_gpio.h"
#include "wifiiot_gpio_ex.h"
#define Pin WIFI_IOT_IO_NAME_GPIO_2
void entry(void)
{
    GpioInit();
    IoSetFunc(Pin, WIFI_IOT_IO_FUNC_GPIO_2_GPIO);
    GpioSetDir(Pin, WIFI_IOT_GPIO_DIR_IN);
    wifiiotGpioValue value;
    sleep(1);
    while(1){
        usleep(500000);

        GpioGetInputVal(Pin, &value);
        if (value == WIFI_IOT_GPIO_VALUE1)
        {
            printf("flame stat:no flame\r\n");
        }else
        {
            printf("flame stat:flame\r\n");
        }
    }
}
APP_FEATURE_INIT(entry);
```

（4）在 case18-flame_detector 中创建模块，构建脚本 BUILD.gn 并初始化模块，代码如下：

```
#//applications/sample/wifi-iot/app/ohos_50/chapter_02/case18-flame_detector/BUILD.gn
static_library("ch_02_flame") {
    sources = [
        "flame_detector_demo.c",
    ]
    include_dirs = [
        "//base/iot_hardware/interfaces/kits/wifiiot_lite",
    ]
}
```

（5）将模块 ch_02_flame 配置到应用子系统，代码如下：

```
#//applications/sample/wifi-iot/app/BUILD.gn
import("//build/lite/config/component/lite_component.gni")
```

```
lite_component("app") {
    features = [
        "ohos_50/chapter_02/case18-flame_detector:ch_02_flame",
    ]
}
```

(6)测试:将火焰传感器与主控板相连,编译应用模块,将固件烧写到开发板,运行IPOP终端工具,以便与开发板相连,复位开发板,将火焰探测头靠近火源,观察IPOP终端工具中的打印信息,如图2-42所示。

```
formatting spiffs...
FileSystem mount ok.
wifi init success!
00 00:00:00 0 196 D 0/HIVIEW: hilog init success.
00 00:00:00 0 196 D 0/HIVIEW: log limit init success.
00 00:00:00 0 196 I 1/SAMGR: Bootstrap core services(count:3).
00 00:00:00 0 196 I 1/SAMGR: Init service:0x4ae9ec TaskPool:0xfa224
00 00:00:00 0 196 I 1/SAMGR: Init service:0x4aea10 TaskPool:0xfa894
00 00:00:00 0 196 I 1/SAMGR: Init service:0x4aeb4c TaskPool:0xfaa54
00 00:00:00 0 228 I 1/SAMGR: Init service 0x4aea10 <time: 0ms> succes
00 00:00:00 0 128 I 1/SAMGR: Init service 0x4ae9ec <time: 0ms> succes
00 00:00:00 0 72 D 0/HIVIEW: hiview init success.
00 00:00:00 0 72 I 1/SAMGR: Init service 0x4aeb4c <time: 0ms> success
00 00:00:00 0 72 I 1/SAMGR: Initialized all core system services!
flame stat:no flame
flame stat:no flame
flame stat:no flame
flame stat:no flame
flame stat:flame
flame stat:flame
flame stat:flame
flame stat:flame
flame stat:flame
```

图 2-42 火焰探测器的运行效果

2.3.10 案例 19:测距仪

本案例为测距仪,实现距离探测功能,其中核心模块为超声波传感器HC-SR04,如图2-43所示。

图 2-43 超声波传感器

上面为超声波发射端与超声波接收端,下面 4 个引脚从左往右依次为 VCC、Trig、Echo、GND。分别与主控板的+5V、GPIO07、GPIO08、GND 相连,其中 GPIO07 为测量指令发送线,GPIO08 为距离检测信号线。

超声波模块是现实生活中常见的模块,该 HC-SR04 模块可以提供 2～400cm 的非接触式距离感测功能,测量精度可以达到 3mm。该模块包含超声波发射器、接收器与控制电路。

超声波传感器 HC-SR04 采用 I/O 口 Trig 触发测距,给最少 10μs 的高电平信号。自动发送 8 个 40kHz 的方波,自动检测是否有信号返回;如果有信号返回,则通过 I/O 口 Echo 输出一个高电平,高电平持续的时间就是超声波从发射到返回的时间。测试距离=(高电平时间×声速)/2,其中声速为 340m/s。

开发步骤如下:

(1) 在 chapter_02 中创建工程目录 case19-range_finder。

(2) 在 case19-range_finder 中创建源码文件 range_finder_demo.c。

(3) 功能实现,代码如下:

```c
//applications/sample/wifi-iot/app/ohos_50/chapter_02/case19-range_finder/range_finder_demo.c
#include <stdio.h>
#include <unistd.h>
#include <ohos_init.h>
#include "hi_time.h"
#include "wifiiot_gpio.h"
#include "wifiiot_gpio_ex.h"
#include "wifiiot_watchdog.h"
void init(void)
{
    GpioInit();
    IoSetFunc(WIFI_IOT_IO_NAME_GPIO_7, WIFI_IOT_IO_FUNC_GPIO_7_GPIO);
    GpioSetDir(WIFI_IOT_IO_NAME_GPIO_7, WIFI_IOT_GPIO_DIR_OUT);
    IoSetFunc(WIFI_IOT_IO_NAME_GPIO_8, WIFI_IOT_IO_FUNC_GPIO_8_GPIO);
    GpioSetDir(WIFI_IOT_IO_NAME_GPIO_8, WIFI_IOT_GPIO_DIR_IN);
}
void get_distance(void)
{
    float distance = 0.0;
    unsigned int flag = 0;
    wifiiotGpioValue value;
    static unsigned long start_time = 0, time = 0;
    //GPIO07 输出高电平,超声波传感器开始测距
    GpioSetOutputVal(WIFI_IOT_IO_NAME_GPIO_7, WIFI_IOT_GPIO_VALUE1);
    usleep(20);
    GpioSetOutputVal(WIFI_IOT_IO_NAME_GPIO_7, WIFI_IOT_GPIO_VALUE0);
```

```c
    while (1)
    {
        GpioGetInputVal(WIFI_IOT_IO_NAME_GPIO_8, &value);
        if (value == WIFI_IOT_GPIO_VALUE1 && flag == 0)
        {
            start_time = hi_get_us();
            flag = 1;
        }
        if (value == WIFI_IOT_GPIO_VALUE0 && flag == 1)
        {
            time = hi_get_us() - start_time;
            start_time = 0;
            break;
        }
    }
    //计算距离
    distance = time * 0.034 / 2;
    printf("\r\ndistance is %.2f cm", distance);
}
void entry(void)
{
    //关闭WatchDog,以后会讲到
    WatchDogDisable();
    //初始化GPIO
    init();
    sleep(1);
    while (1)
    {
        //获取距离并打印
        get_distance();
        sleep(1);
    }
}
APP_FEATURE_INIT(entry);
```

（4）在case19-range_finder中创建模块，构建脚本BUILD.gn并初始化模块，代码如下：

```
#//applications/sample/wifi-iot/app/ohos_50/chapter_02/case19-range_finder/BUILD.gn
static_library("ch_02_range_finder") {
    sources = [
        "range_finder_demo.c",
    ]
    include_dirs = [
```

```
        "//base/iot_hardware/interfaces/kits/wifiiot_lite",
    ]
}
```

（5）将模块 ch_02_range_finder 配置到应用子系统，代码如下：

```
#//applications/sample/wifi-iot/app/BUILD.gn
import("//build/lite/config/component/lite_component.gni")
lite_component("app") {
    features = [
        "ohos_50/chapter_02/case19-range_finder:ch_02_range_finder",
    ]
}
```

（6）测试：将超声波传感器与主控板相连，编译应用模块，将固件烧写到开发板，运行 IPOP 终端工具，以便与开发板相连，复位开发板，观察 IPOP 终端工具中的打印信息，如图 2-44 所示。

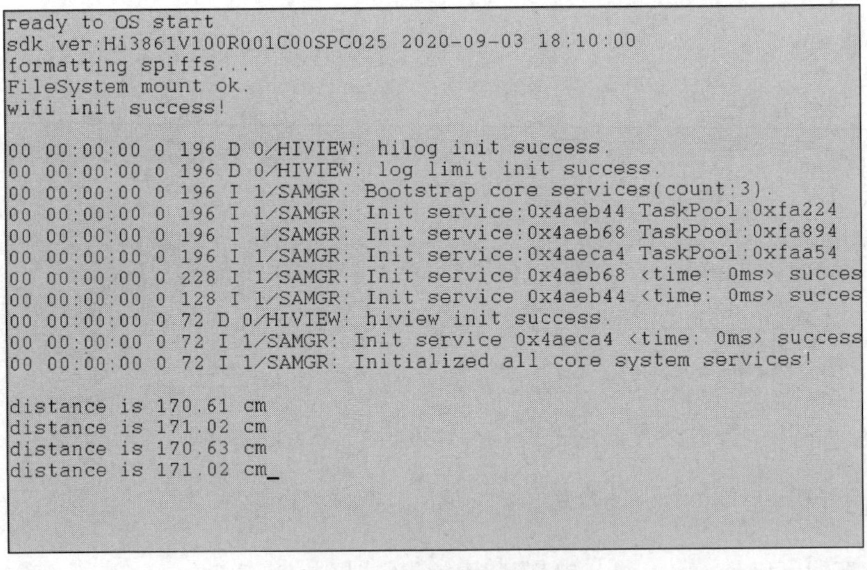

```
ready to OS start
sdk ver:Hi3861V100R001C00SPC025 2020-09-03 18:10:00
formatting spiffs...
FileSystem mount ok.
wifi init success!

00 00:00:00 0 196 D 0/HIVIEW: hilog init success.
00 00:00:00 0 196 D 0/HIVIEW: log limit init success.
00 00:00:00 0 196 I 1/SAMGR: Bootstrap core services(count:3).
00 00:00:00 0 196 I 1/SAMGR: Init service:0x4aeb44 TaskPool:0xfa224
00 00:00:00 0 196 I 1/SAMGR: Init service:0x4aeb68 TaskPool:0xfa894
00 00:00:00 0 196 I 1/SAMGR: Init service:0x4aeca4 TaskPool:0xfaa54
00 00:00:00 0 228 I 1/SAMGR: Init service 0x4aeb68 <time: 0ms> succes
00 00:00:00 0 128 I 1/SAMGR: Init service 0x4aeb44 <time: 0ms> succes
00 00:00:00 0 72 D 0/HIVIEW: hiview init success.
00 00:00:00 0 72 I 1/SAMGR: Init service 0x4aeca4 <time: 0ms> success
00 00:00:00 0 72 I 1/SAMGR: Initialized all core system services!

distance is 170.61 cm
distance is 171.02 cm
distance is 170.63 cm
distance is 171.02 cm_
```

图 2-44　测距仪的运行效果

2.4　PWM

多种脉冲宽度调制（Pulse Width Modulation，PWM）用于输出波形，共有 6 个端口，调用 PWM 端口时需要进行 I/O 复用，从而实现在指定端口输出指定占空比的方波，可应用于小车、轮船模型上，用于控制电机的转动速度，有关 PWM 的枚举、函数声明在头文件 wifiiot_ pwm.h 中。

Hi3861 芯片的 PWM 设备接口的开发流程如下：
（1）在//vendor/hisi/hi3861/hi3861/build/config/usr_config.mk 文件中开启 PWM。
（2）分析电路，确定 PWM 端口。
（3）调用函数 GpioInit 初始化 GPIO 引脚。
（4）将引脚功能设置为 PWM。
（5）调用函数 PwmInit 初始化 PWM 端口。
（6）调用函数 PwmStart 输出方波。
（7）调用函数 PwmStop 停止输出。

PwmStart 的原型如下：

```
//base/iot_hardware/interfaces/kits/wifiiot_lite/wifiiot_pwm.h
/**
 * @brief 基于输入参数输出 PWM 信号
 *
 * 此功能基于以下参数从指定端口输出 PWM 信号
 * 配置的分频倍数和占空比
 *
 * @param 端口指示 PWM 端口号
 * @param 占空比表示 PWM 占空比
 * @param freq 表示分频倍数
 * @return 如果操作成功,则返回{@link WIFI_IOT_SUCCESS}
 * 返回{@link wifiiot_errno.h}中定义的错误代码,如果为否,则返回
 * @since 1.0
 * @version 1.0
 */
unsigned int PwmStart(wifiiotPwmPort port, unsigned short duty, unsigned short freq);
```

PwmStop 的原型如下：

```
//base/iot_hardware/interfaces/kits/wifiiot_lite/wifiiot_pwm.h
/**
 * @brief 停止 PWM 信号输出
 *
 * @param 端口指示 PWM 端口号
 * @return 如果操作成功,则返回{@link WIFI_IOT_SUCCESS}
 * 返回{@link wifiiot_errno.h}中定义的错误代码,如果为否,则返回
 * @since 1.0
 * @version 1.0
 */
unsigned int PwmStop(wifiiotPwmPort port);
```

枚举 wifiiotPwmClkSource 的原型如下：

```
//base/iot_hardware/interfaces/kits/wifiiot_lite/wifiiot_pwm.h
/**
 * @brief Enumerates PWM clock sources.
 */
typedef enum {
    /** 160 MHz working clock */
    WIFI_IOT_PWM_CLK_160M,
    /** 24 MHz or 40 MHz external crystal */
    WIFI_IOT_PWM_CLK_XTAL,
    /** Maximum value */
    WIFI_IOT_PWM_CLK_MAX
} wifiiotPwmClkSource;
```

枚举 wifiiotPwmPort 的原型如下:

```
//base/iot_hardware/interfaces/kits/wifiiot_lite/wifiiot_pwm.h
/**
 * @brief Enumerates PWM ports.
 */
typedef enum {
    /** PWM0 */
    WIFI_IOT_PWM_PORT_PWM0 = 0,
    /** PWM1 */
    WIFI_IOT_PWM_PORT_PWM1 = 1,
    /** PWM2 */
    WIFI_IOT_PWM_PORT_PWM2 = 2,
    /** PWM3 */
    WIFI_IOT_PWM_PORT_PWM3 = 3,
    /** PWM4 */
    WIFI_IOT_PWM_PORT_PWM4 = 4,
    /** PWM5 */
    WIFI_IOT_PWM_PORT_PWM5 = 5,
    /** Maximum value */
    WIFI_IOT_PWM_PORT_MAX
} wifiiotPwmPort;
```

2.4.1 案例20:自动门

本案例为自动门,实现自动开门和关门功能。其核心模块为直流电机驱动控制模块,如图2-45所示。

图中左面为可正反转直流电机,右面为电机驱动板。将直流电机电源线插在直流电机驱动板左上角的插槽,将主控板插到母板上,再将电机驱动板插到母板上,如图2-46所示。

本案例是通过输出两路PWM信号来完成电机的正反转控制的,其中一路PWM信号控制正转,另一路控制反转,两路不可以同时输出。具体可参考电机驱动控制电路图及控制

图 2-45 减速电机与机器人板

图 2-46 组装后的开发板

芯片手册。

开发步骤如下:

(1) 在//vendor/hisi/hi3861/hi3861/build/config/usr_config.mk 文件中开启 PWM。打开//vendor/hisi/hi3861/hi3861/build/config/usr_config.mk 文件。在 usr_config.mk 文件中找到 #CONFIG_PWM_SUPPORT is not set。在 #CONFIG_PWM_SUPPORT is not set 的下面插入一条语句开启 PWM,代码如下:

```
#//vendor/hisi/hi3861/hi3861/build/config/usr_config.mk
CONFIG_PWM_SUPPORT = y
```

(2) 在 chapter_02 中创建工程目录 case20-automatic_door。
(3) 在 case20-automatic_door 中创建源码文件 automatic_door_demo.c。
(4) 功能实现,代码如下:

```
//applications/sample/wifi-iot/app/ohos_50/chapter_02/case20-automatic_door/automatic_door_demo.c
#include <stdio.h>
#include <unistd.h>
```

```c
#include <ohos_init.h>
#include "wifiiot_pwm.h"
#include "wifiiot_gpio.h"
#include "wifiiot_gpio_ex.h"
//初始化 PWM
void pwm_init(void)
{
    GpioInit();
    //将 GPIO0 引脚功能设置为 PWM3 输出
    IoSetFunc(WIFI_IOT_IO_NAME_GPIO_0, WIFI_IOT_IO_FUNC_GPIO_0_PWM3_OUT);
    //将 GPIO1 引脚功能设置为 PWM4 输出
    IoSetFunc(WIFI_IOT_IO_NAME_GPIO_1, WIFI_IOT_IO_FUNC_GPIO_1_PWM4_OUT);
    //初始化 PWM3、PWM4
    PwmInit(WIFI_IOT_PWM_PORT_PWM3);
    PwmInit(WIFI_IOT_PWM_PORT_PWM4);
}
void entry(void)
{
    pwm_init();
    while (1)
    {
        printf("open door\r\n");
        //输出 PWM3 信号,实现开门功能
        PwmStart(WIFI_IOT_PWM_PORT_PWM3, 3000, 3000);
        sleep(3);
        //停止 PWM3 信号输出,电机停止转动
        PwmStop(WIFI_IOT_PWM_PORT_PWM3);
        sleep(2);
        printf("close door\r\n");
        //输出 PWM4 信号,实现关门功能
        PwmStart(WIFI_IOT_PWM_PORT_PWM4, 3000, 3000);
        sleep(3);
        //停止 PWM4 信号输出,电机停止转动
        PwmStop(WIFI_IOT_PWM_PORT_PWM4);
        sleep(2);
    }
}
APP_FEATURE_INIT(entry);
```

(5) 在 case20-automatic_door 中创建模块,构建脚本 BUILD.gn 并初始化模块,代码如下:

```
#//applications/sample/wifi-iot/app/ohos_50/chapter_02/case20-automatic_door/BUILD.gn
static_library("ch_02_automatic_door") {
    sources = [
```

```
        "automatic_door_demo.c",
    ]
    include_dirs = [
        "//base/iot_hardware/interfaces/kits/wifiiot_lite",
    ]
}
```

(6) 将模块 ch_02_automatic_door 配置到应用子系统,代码如下:

```
#//applications/sample/wifi-iot/app/BUILD.gn
import("//build/lite/config/component/lite_component.gni")
lite_component("app") {
    features = [
        "ohos_50/chapter_02/case20-automatic_door:ch_02_automatic_door",
    ]
}
```

(7) 测试:将直流电机与直流电机驱动板相连,将电机驱动板插到母板上,将主控板插到母板上,编译应用模块,将固件烧写到开发板,复位开发板,直流电机如图 2-47 所示。

图 2-47　减速电机的运行效果

注意:如果在编译的最后报未声明与 PWM 相关的方法的错误,则应在//vendor/hisi/hi3861/hi3861/build/config/usr_config.mk 文件中开启 PWM。

2.4.2　案例 21:炫彩灯

本案例为炫彩灯,实现多种颜色随机显示功能。核心模块为高亮红绿蓝(RGB)三色光 LED 模块,如图 2-48 所示。

将主控板插到母板上,再将人体感应板插到母板上,如图 2-49 所示。

本案例的实现原理为红黄蓝(RGB)三原色原理。通过控制 RGB 3 种基本颜色的亮度来混合出不同颜色的光。

开发步骤如下:

(1) 在//vendor/hisi/hi3861/hi3861/build/config/usr_config.mk 文件中开启 PWM。

(2) 在 chapter_02 中创建工程目录 case21-colorful_light。

(3) 在 case21-colorful_light 中创建源码文件 colorful_light_demo.c。

(4) 功能实现,代码如下:

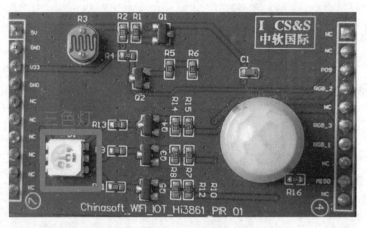

图 2-48 三色灯

图 2-49 组装后的开发板

```
//applications/sample/wifi-iot/app/ohos_50/chapter_02/case21-colorful_light/colorful_light_demo.c
#include <stdio.h>
#include <unistd.h>
#include <stdlib.h>
#include <ohos_init.h>
#include "wifiiot_gpio.h"
#include "wifiiot_gpio_ex.h"
#include "wifiiot_pwm.h"

#define RED_LED_PIN_NAME WIFI_IOT_IO_NAME_GPIO_10
#define GREEN_LED_PIN_NAME WIFI_IOT_IO_NAME_GPIO_11
```

```c
#define BLUE_LED_PIN_NAME WIFI_IOT_IO_NAME_GPIO_12
#define PWM_DUTY 64000
#define PWM_FREQ_DIVISION 32767
void init(void)
{
    GpioInit();
    IoSetFunc(RED_LED_PIN_NAME, WIFI_IOT_IO_FUNC_GPIO_10_PWM1_OUT);
    IoSetFunc(GREEN_LED_PIN_NAME, WIFI_IOT_IO_FUNC_GPIO_11_PWM2_OUT);
    IoSetFunc(BLUE_LED_PIN_NAME, WIFI_IOT_IO_FUNC_GPIO_12_PWM3_OUT);
    PwmInit(WIFI_IOT_PWM_PORT_PWM1); //R
    PwmInit(WIFI_IOT_PWM_PORT_PWM2); //G
    PwmInit(WIFI_IOT_PWM_PORT_PWM3); //B
}
void entry(void)
{
    init();
    while (1)
    {
        PwmStart(WIFI_IOT_PWM_PORT_PWM1, rand(), PWM_FREQ_DIVISION);
        PwmStart(WIFI_IOT_PWM_PORT_PWM2, rand(), PWM_FREQ_DIVISION);
        PwmStart(WIFI_IOT_PWM_PORT_PWM3, rand(), PWM_FREQ_DIVISION);
        usleep(300000);
    }
}
APP_FEATURE_INIT(entry);
```

（5）在 case21-colorful_light 中创建模块，构建脚本 BUILD.gn 并初始化模块，代码如下：

```
#//applications/sample/wifi-iot/app/ohos_50/chapter_02/case21-colorful_light/BUILD.gn
static_library("ch_02_colorful_light") {
    sources = [
        "colorful_light_demo.c",
    ]
    include_dirs = [
        "//base/iot_hardware/interfaces/kits/wifiiot_lite",
    ]
}
```

（6）将模块 ch_02_colorful_light 配置到应用子系统，代码如下：

```
#//applications/sample/wifi-iot/app/BUILD.gn
import("//build/lite/config/component/lite_component.gni")
lite_component("app") {
    features = [
```

```
            "ohos_50/chapter_02/case21-colorful_light:ch_02_colorful_light",
    ]
}
```

(7) 测试：将人体传感器插到母板上，将主控板插到母板上，编译应用模块，将固件烧写到开发板，复位开发板，观察三色灯的颜色与亮度，如图 2-50 所示。

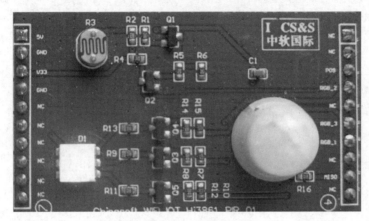

图 2-50　炫彩灯的运行效果

2.4.3　案例 22：救护车警报器

本案例为救护车警报器，实现发出救护车警报声音功能，核心模块为蜂鸣器，如图 2-51 所示。

图 2-51　蜂鸣器

右面的 3 个引脚从上往下依次为 GND、I/O、VCC，分别与将主控板的 GND、GPIO09（PWM0）、+5V 相连。

本案例的实现原理为循环地向蜂鸣器输出两个不同频率的 PWM 方波，使用蜂鸣器发出两种交替变化的声音。

开发步骤如下：

(1) 在 //vendor/hisi/hi3861/hi3861/build/config/usr_config.mk 文件中开启 PWM。

(2) 在 chapter_02 中创建工程目录 case22-ambulance_alarm。

(3) 在 case22-ambulance_alarm 中创建源码文件 ambulance_alarm_demo.c。

(4) 功能实现，代码如下：

```c
//applications/sample/wifi-iot/app/ohos_50/chapter_02/case22-ambulance_alarm/ambulance_alarm_demo.c
#include <stdio.h>
#include <unistd.h>
#include <ohos_init.h>
#include "wifiiot_pwm.h"
#include "wifiiot_gpio.h"
#include "wifiiot_gpio_ex.h"
void init_pwm(void)
{
    GpioInit();
    IoSetFunc(WIFI_IOT_IO_NAME_GPIO_9, WIFI_IOT_IO_FUNC_GPIO_9_PWM0_OUT);
    PwmInit(WIFI_IOT_PWM_PORT_PWM0);
}
int pu[2] = {
    20249,          //对应 C 调 7
    38223           //对应 C 调 1
};
void entry(void)
{
    init_pwm();

    while (1)
    {
        for (int i = 0; i < 2; i++)
        {
            PwmStart(WIFI_IOT_PWM_PORT_PWM0, pu[i] / 2, pu[i]);
            sleep(1);
        }
    }
}
APP_FEATURE_INIT(entry);
```

(5) 在 case22-ambulance_alarm 中创建模块, 构建脚本 BUILD.gn 并初始化模块, 代码如下:

```
#//applications/sample/wifi-iot/app/ohos_50/chapter_02/case22-ambulance_alarm/BUILD.gn
static_library("ch_02_ambulance_alarm") {
    sources = [
        "ambulance_alarm_demo.c",
    ]
    include_dirs = [
        "//base/iot_hardware/interfaces/kits/wifiiot_lite",
    ]
}
```

(6) 将模块 ch_02_ambulance_alarm 配置到应用子系统,代码如下:

```
#//applications/sample/wifi-iot/app/BUILD.gn
import("//build/lite/config/component/lite_component.gni")
lite_component("app") {
  features = [
    "ohos_50/chapter_02/case22-ambulance_alarm:ch_02_ambulance_alarm",
  ]
}
```

(7) 测试:将蜂鸣器与主控板相连,编译应用模块,将固件烧写到开发板,复位开发板,聆听蜂鸣器发出的救护车警报。

2.4.4 案例23:音乐盒

本案例为音乐盒,核心模块为蜂鸣器,如图 2-52 所示。

图 2-52 蜂鸣器

案例是通过控制 PWM 输出不同方波频率实现音符的输出,通过延时来完成指定节拍。开发步骤如下:

(1) 在//vendor/hisi/hi3861/hi3861/build/config/usr_config.mk 文件中开启 PWM。
(2) 在 chapter_02 中创建工程目录 case23-music_box。
(3) 在 case23-music_box 中创建源码文件 music_box_demo.c。
(4) 功能实现,代码如下:

```
//applications/sample/wifi-iot/app/ohos_50/chapter_02/case23-music_box/music_box_demo.c
#include <stdio.h>
#include <unistd.h>
#include <ohos_init.h>
#include "wifiiot_pwm.h"
#include "wifiiot_gpio.h"
#include "wifiiot_gpio_ex.h"
#include "wifiiot_watchdog.h"
typedef unsigned short uint16_t;
typedef unsigned char uint8_t;
typedef unsigned int uint32_t;
static const uint16_t g_tuneFreqs[] = {
```

```
    0,                      //40MHz 对应的分频系数
    38223, //1046.5
    34052, //1174.7
    30338, //1318.5
    28635, //1396.9
    25511, //1568
    22728, //1760
    20249, //1975.5
    51021 //5_783.99      //第 1 个八度的 5
};

//曲谱音符,《两只老虎》简谱的网址为 http://www.jianpu.cn/pu/33/33945.htm
static const uint8_t g_scoreNotes[] = {
    1, 2, 3, 1, 1, 2, 3, 1,
    3, 4, 5, 3, 4, 5,
    5, 6, 5, 4, 3, 1,
    5, 6, 5, 4, 3, 1,
    1, 8, 1,
    1, 8, 1,
};
//曲谱时值
static const uint8_t g_scoreDurations[] = {
    4, 4, 4, 4, 4, 4, 4, 4,
    4, 4, 8, 4, 4, 8,
    3, 1, 3, 1, 4, 4,
    3, 1, 3, 1, 4, 4,
    4, 4, 8,
    4, 4, 8,
};
void entry(void)
{
    GpioInit();
    //将蜂鸣器引脚设置为 PWM 功能
    IoSetFunc(WIFI_IOT_IO_NAME_GPIO_9, WIFI_IOT_IO_FUNC_GPIO_9_PWM0_OUT);
    PwmInit(WIFI_IOT_PWM_PORT_PWM0);
    WatchDogDisable();
    sleep(1);
    printf("Music_box_demo start!\r\n");
    while (1)
    {
        for (size_t i = 0; i < sizeof(g_scoreNotes) / sizeof(g_scoreNotes[0]); i++)
        {
            //音符
            uint32_t tune = g_scoreNotes[i];
            //获取频率
            uint16_t freqDivisor = g_tuneFreqs[tune];
```

```c
            //音符时间,节拍
            uint32_t tuneInterval = g_scoreDurations[i] * (125 * 1000);

            //打印音符的频率、时长(μs)等相关信息
            printf("%d %d %d %d\r\n", tune, (40 * 1000 * 1000) / freqDivisor, freqDivisor,
tuneInterval);

            PwmStart(WIFI_IOT_PWM_PORT_PWM0, freqDivisor / 100, freqDivisor);
            usleep(tuneInterval); //持续响 tuneInterval 设定的时间
            PwmStop(WIFI_IOT_PWM_PORT_PWM0);
        }
        usleep(2000 * 1000);
    }
}
APP_FEATURE_INIT(entry);
```

(5) 在 case23-music_box 中创建模块,构建脚本 BUILD.gn 并初始化模块,代码如下:

```
#//applications/sample/wifi-iot/app/ohos_50/chapter_02/case23-music_box/BUILD.gn
static_library("ch_02_music_box") {
    sources = [
        "music_box_demo.c",
    ]
    include_dirs = [
        "//base/iot_hardware/interfaces/kits/wifiiot_lite",
    ]
}
```

(6) 将模块 ch_02_music_box 配置到应用子系统,代码如下:

```
#//applications/sample/wifi-iot/app/BUILD.gn
import("//build/lite/config/component/lite_component.gni")
lite_component("app") {
    features = [
        "ohos_50/chapter_02/case24-music_box:ch_02_music_box",
    ]
}
```

(7) 测试:将蜂鸣器与主控板相连,编译应用模块,将固件烧写到开发板,复位开发板,聆听蜂鸣器发出的音乐《两只老虎》,并观察 IPOP 终端工具输出的音调信息,如图 2-53 所示。

```
ready to OS start
sdk ver:Hi3861V100R001C00SPC025 2020-09-03 18:10:00
FileSystem mount ok.
wifi init success!

00 00:00:00 0 196 D 0/HIVIEW: hilog init success.
00 00:00:00 0 196 D 0/HIVIEW: log limit init success.
00 00:00:00 0 196 I 1/SAMGR: Bootstrap core services(count:3).
00 00:00:00 0 196 I 1/SAMGR: Init service:0x4aec2c TaskPool:0xfa224
00 00:00:00 0 196 I 1/SAMGR: Init service:0x4aec50 TaskPool:0xfa894
00 00:00:00 0 196 I 1/SAMGR: Init service:0x4aeddc TaskPool:0xfaa54
00 00:00:00 0 228 I 1/SAMGR: Init service 0x4aec50 <time: 0ms> succes
00 00:00:00 0 128 I 1/SAMGR: Init service 0x4aec2c <time: 0ms> succes
00 00:00:00 0 72 D 0/HIVIEW: hiview init success.
00 00:00:00 0 72 I 1/SAMGR: Init service 0x4aeddc <time: 0ms> success
00 00:00:00 0 72 I 1/SAMGR: Initialized all core system services!
Music_box_demo start!
1 1046 38223 500000
2 1174 34052 500000
3 1318 30338 500000
1 1046 38223 500000
```

图 2-53 音乐盒的运行效果

2.5 I²C

集成电路总线(Inter-Integrated Circuit，I²C)用于低速设备之间的通信，一共只有两条线：一条是双向的串行数据线 SDA，另一条是串行时钟线 SCL。

(1) SDA(Serial Data)是数据线，用来传输数据。

(2) SCL(Serial Clock Line)是时钟线，用来控制数据发送的时序。

它是一种串行通信总线，使用多主从架构，是由飞利浦公司在 20 世纪 80 年代初设计的，方便了主板、嵌入式系统或手机与周边设备组件之间的通信。由于其简单，它被广泛用于微控制器与传感器阵列、显示器、IoT 设备、EEPROM 等之间的通信。

它共有两个端口，调用 I²C 端口时需要进行 IO 复用，从而可以在指定端口接收或发送数据，有关 I²C 的枚举、函数声明在头文件 wifiiot_i2c.h 中。

Hi3861 芯片的 I²C 设备接口的开发流程如下：

(1) 在 //vendor/hisi/hi3861/hi3861/build/config/usr_config.mk 文件中开启 I²C。

(2) 分析电路，确定 I²C 端口。

(3) 调用函数 GpioInit 初始化 GPIO 引脚。

(4) 将两个引脚功能分别设置为 I2C_SDA 与 I2C_SCL。

(5) 调用函数 I2cInit 初始化 I²C 波特率。

(6) 调用函数 I2cWrite 发送数据(可选)。

(7) 调用函数 I2cRead 接收数据(可选)。

函数 I2cInit 的原型如下：

```
//base/iot_hardware/interfaces/kits/wifiiot_lite/wifiiot_i2c.h
/**
 * @brief 以指定的波特率初始化 I²C 设备
 *
 *
 *
 * @param id 表示 I²C 设备的 id
 * @param - baudrate 表示 I²C 的波特率
 * @return 如果操作成功,则返回{@link WIFI_IOT_SUCCESS}
 * @返回{@link wifiiot_errno.h}中定义的错误代码,如果为否,则返回
 * @since 1.0
 * @version 1.0
 */
unsigned int I2cInit(wifiiotI2cIdx id, unsigned int baudrate);
```

函数 I2cWrite 的原型如下:

```
//base/iot_hardware/interfaces/kits/wifiiot_lite/wifiiot_i2c.h
/**
 * @brief 将数据写入 I²C 设备
 *
 *
 *
 * @param id 表示 I²C 设备 id
 * @param deviceAddr 表示 I²C 的设备地址
 * @param i2cData 表示指向要写入的数据描述符的指针
 * @return 如果操作成功,则返回{@link WIFI_IOT_SUCCESS}
 * returns 在{@link wifiiot_errno.h}中定义的错误代码
 * @since 1.0
 * @version 1.0
 */
unsigned int I2cWrite(wifiiotI2cIdx id, unsigned short deviceAddr, const wifiiotI2cData *
i2cData);
```

函数 I2cRead 的原型如下:

```
//base/iot_hardware/interfaces/kits/wifiiot_lite/wifiiot_i2c.h
/**
 * @brief 从 I²C 设备读取数据
 *
 * 读取的数据将保存到<b> i2cData </b>指定的地址
 *
 * @param id 表示 I²C 的设备 id
 * @param deviceAddr 表示 I²C 的设备地址
```

```
 * @param i2cData 表示指向要读取的数据描述符的指针
 * @return 如果操作成功,则返回{@link WIFI_IOT_SUCCESS}
 * returns 在{@link wifiiot_errno.h}中定义的错误代码
 * @since 1.0
 * @version 1.0
 */
unsigned int I2cRead(wifiiotI2cIdx id, unsigned short deviceAddr, const wifiiotI2cData *
i2cData);
```

枚举 wifiiotI2cData 的原型如下:

```
//base/iot_hardware/interfaces/kits/wifiiot_lite/wifiiot_i2c.h
/**
 * @brief Defines I2C data transmission attributes.
 */
typedef struct {
    /** Pointer to the buffer storing data to send */
    unsigned char * sendBuf;
    /** Length of data to send */
    unsigned int sendLen;
    /** Pointer to the buffer for storing data to receive */
    unsigned char * receiveBuf;
    /** Length of data received */
    unsigned int receiveLen;
} wifiiotI2cData;
```

枚举 wifiiotI2cIdx 的原型如下:

```
//base/iot_hardware/interfaces/kits/wifiiot_lite/wifiiot_i2c.h
/**
 * @brief Enumerates I²C hardware indexes
 */
typedef enum {
    /** I²C hardware index 0 */
    WIFI_IOT_I2C_IDX_0,
    /** I²C hardware index 1 */
    WIFI_IOT_I2C_IDX_1,
} wifiiotI2cIdx;
```

2.5.1 案例24:温湿度监测仪

本案例为温湿度监测仪,实现环境温湿度监测功能。核心模块为SHT40温湿度传感器,如图2-54所示。

将主控板插到母板上,再将温湿度感应板插到母板上,如图2-55所示。

图 2-54　SHT40 温湿度传感器

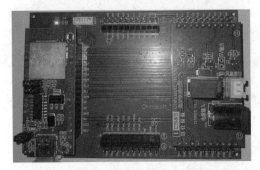

图 2-55　温湿度监测仪组装后的开发板

案例实现思路：通过 I^2C 向 SHT40 芯片发送监测指令，然后读取返回的数据，再解析数据，最后输出温湿度数据。

开发步骤如下：

（1）在 //vendor/hisi/hi3861/hi3861/build/config/usr_config.mk 文件中开启 I^2C。打开 //vendor/hisi/hi3861/hi3861/build/config/usr_config.mk 文件。在 usr_config.mk 中找到 #CONFIG_I2C_SUPPORT is not set。在 #CONFIG_I2C_SUPPORT is not set 的下面插入一条语句，以便开启 I^2C，代码如下：

```
#//vendor/hisi/hi3861/hi3861/build/config/usr_config.mk
CONFIG_I2C_SUPPORT = y
```

（2）在 chapter_02 中创建工程目录 case24-temp_and_hum_detector。

（3）在 case24-temp_and_hum_detector 中创建源码文件 temp_and_hum_detector_demo.c。

（4）引用必要的头文件，声明必要的宏及变量，代码如下：

```
#include <stdio.h>
#include <unistd.h>
#include <string.h>
```

```c
#include "ohos_init.h"

#include "wifiiot_gpio.h"
#include "wifiiot_gpio_ex.h"
#include "wifiiot_i2c.h"

#define SHT40_I2C_IDX 0                      //I²C的设备号
#define SHT40_I2C_BAUDRATE (400 * 1000)      //I²C的波特率
#define SHT40_ADDR 0x44                      //SHT40的设备地址
#define SHT40_STATUS_RESPONSE_MAX 6          //读取传感器数据长度
#define SHT40_CMD_TRIGGER 0xFD               //测量命令
typedef unsigned int uint32_t;
typedef unsigned char uint8_t;
int temp = 0;
int hum = 0;
```

(5) 创建函数 sht40_init,初始化 I²C,代码如下:

```c
//applications/sample/wifi-iot/app/ohos_50/chapter_02/case24-temp_and_hum_detector/temp_and_hum_detector_demo.c
//SHT40 I²C初始化
void sht40_init(void)
{
    GpioInit();
    IoSetFunc(WIFI_IOT_IO_NAME_GPIO_13, WIFI_IOT_IO_FUNC_GPIO_13_I2C0_SDA);
    IoSetFunc(WIFI_IOT_IO_NAME_GPIO_14, WIFI_IOT_IO_FUNC_GPIO_14_I2C0_SCL);

    I2cInit(SHT40_I2C_IDX, SHT40_I2C_BAUDRATE);     //初始化I2C波特率
}
```

(6) 创建函数 sht40_write 实现通过 I²C 总线将数据发送到 SHT40 温湿度传感器功能,代码如下:

```c
//applications/sample/wifi-iot/app/ohos_50/chapter_02/case24-temp_and_hum_detector/temp_and_hum_detector_demo.c
//将数据写到 SHT40
static uint32_t sht40_write(uint8_t *buffer, uint32_t buffLen)
{
    wifiiotI2cData i2cData = {buffer, buffLen, NULL, 0};
    uint32_t retval = I2cWrite(SHT40_I2C_IDX, (SHT40_ADDR << 1) | 0, &i2cData);
    if (retval != 0)
    {
        printf("I2cWrite(%02X) failed, %0X!\n", buffer[0], retval);
        return retval;
    }
    return 0;
}
```

(7) 创建函数 sht40_read,实现通过 I²C 总线读取 SHT40 温湿度传感器发送的数据,代码如下:

```c
//applications/sample/wifi-iot/app/ohos_50/chapter_02/case24-temp_and_hum_detector/temp_and_hum_detector_demo.c
//从 SHT40 中读取数据
static uint32_t sht40_read(uint8_t * buffer, uint32_t buffLen)
{
    uint32_t retval;
    wifiiotI2cData i2cData = {NULL, 0, buffer, buffLen};
    retval = I2cRead(SHT40_I2C_IDX, (SHT40_ADDR << 1) | 1, &i2cData);
    if (retval != 0)
    {
        printf("I2cRead() failed, %0X!\n", retval);
        return retval;
    }
    return 0;
}
```

(8) 创建函数 sht40_start_measure 与 sht40_get_measure_result,实现测量并获取温湿度功能,代码如下:

```c
//applications/sample/wifi-iot/app/ohos_50/chapter_02/case24-temp_and_hum_detector/temp_and_hum_detector_demo.c
//测量
uint32_t sht40_start_measure(void)
{
    uint8_t triggerCmd[] = {SHT40_CMD_TRIGGER};
    return sht40_write(triggerCmd, sizeof(triggerCmd));
}

//获取测量结果
uint32_t SHT40_sht40_get_measure_result(int * temp, int * humi)
{
    uint32_t retval = 0;
    //用来存放表示温度的数据
    float t_degC = 0.0;
    //用来存放表示湿度的数据
    float rh_pRH = 0.0;
    float t_ticks = 0.0;
    float rh_ticks = 0.0;
    if (temp == NULL || humi == NULL)
    {
        return -1;
    }
    uint8_t buffer[SHT40_STATUS_RESPONSE_MAX] = {0};
    //以 0x0 初始化内存
    memset(buffer, 0x0, sizeof(buffer));
```

```c
//获取测量结果
retval = sht40_read(buffer, sizeof(buffer));
if (retval != 0)
{
    return retval;
}
t_ticks = buffer[0] * 256 + buffer[1];
rh_ticks = buffer[3] * 256 + buffer[4];
//计算温湿度
t_degC = -45 + 175 * t_ticks / 65535;
rh_pRH = -6 + 125 * rh_ticks / 65535;
if (rh_pRH >= 100)
{
    rh_pRH = 100;
}
if (rh_pRH < 0)
{
    rh_pRH = 0;
}
*humi = (uint8_t)rh_pRH;
*temp = (uint8_t)t_degC;
return 0;
}
```

(9) 创建应用模块入口函数 entry,实现循环测量打印温湿度功能,代码如下:

```c
//applications/sample/wifi-iot/app/ohos_50/chapter_02/case24-temp_and_hum_detector/temp_and_hum_detector_demo.c
//主功能函数,实现循环测量打印温湿度功能
void entry(void)
{
    sleep(1);                              //睡眠
    sht40_init();                          //温湿度传感器IO初始化
    while (1)
    {
        sht40_start_measure();
        usleep(20 * 1000);
        sht40_get_measure_result(&temp, &hum);   //获取当前温湿度值
        printf("temp:%4d,hum:%4d\r\n", temp, hum);
        sleep(1);                          //睡眠
    }
}
```

(10) 调用宏 APP_FEATURE_INIT,将模块入口函数初始化为 entry,代码如下:

```c
APP_FEATURE_INIT(entry);
```

(11) 在 case24-temp_and_hum_detector 中创建模块,构建脚本 BUILD.gn 并初始化模块,代码如下:

```
#//applications/sample/wifi-iot/app/ohos_50/chapter_02/case24-temp_and_hum_detector/BUILD.gn
static_library("ch_02_temp_and_hum_detector") {
    sources = [
        "temp_and_hum_detector_demo.c",
    ]

    include_dirs = [
        "//base/iot_hardware/interfaces/kits/wifiiot_lite",
    ]
}
```

(12) 将模块 ch_02_temp_and_hum_detector 配置到应用子系统,代码如下:

```
#//applications/sample/wifi_iot/app/BUILD.gn
import("//build/lite/config/component/lite_component.gni")
lite_component("app") {
    features = [
        "ohos_50/chapter_02/case24-temp_and_hum_detector:ch_02_temp_and_hum_detector",
    ]
}
```

(13) 测试:将温湿度感应板插到母板上,将主控板插到母板上,编译应用模块,将固件烧写到开发板,运行 IPOP 终端工具,以便与开发板相连,复位开发板,观察 IPOP 终端工具的打印信息,如图 2-56 所示。

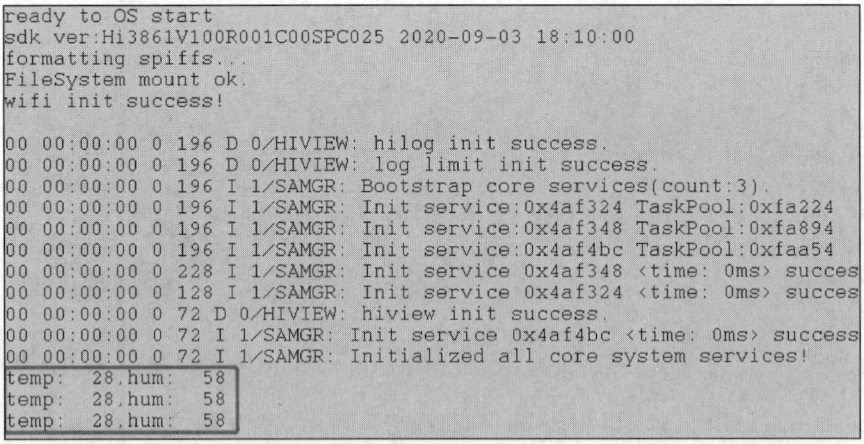

```
ready to OS start
sdk ver:Hi3861V100R001C00SPC025 2020-09-03 18:10:00
formatting spiffs...
FileSystem mount ok.
wifi init success!

00 00:00:00 0 196 D 0/HIVIEW: hilog init success.
00 00:00:00 0 196 D 0/HIVIEW: log limit init success.
00 00:00:00 0 196 I 1/SAMGR: Bootstrap core services(count:3).
00 00:00:00 0 196 I 1/SAMGR: Init service:0x4af324 TaskPool:0xfa224
00 00:00:00 0 196 I 1/SAMGR: Init service:0x4af348 TaskPool:0xfa894
00 00:00:00 0 196 I 1/SAMGR: Init service:0x4af4bc TaskPool:0xfaa54
00 00:00:00 0 228 I 1/SAMGR: Init service 0x4af348 <time: 0ms> succes
00 00:00:00 0 128 I 1/SAMGR: Init service 0x4af324 <time: 0ms> succes
00 00:00:00 0 72 D 0/HIVIEW: hiview init success.
00 00:00:00 0 72 I 1/SAMGR: Init service 0x4af4bc <time: 0ms> success
00 00:00:00 0 72 I 1/SAMGR: Initialized all core system services!
temp:  28,hum:  58
temp:  28,hum:  58
temp:  28,hum:  58
```

图 2-56 温湿度监测仪的运行效果

注意：如果在编译的最后报未声明与 I^2C 相关的方法的错误，则应在//vendor/hisi/hi3861/hi3861/build/config/usr_config.mk 文件中开启 I^2C。

2.5.2 案例 25：电子阅读器

本案例为电子阅读器，实现文本显示功能，核心模块为 OLED 显示屏，如图 2-57 所示。

图 2-57 OLED 显示屏

将主控板插到母板上，再将 LCD 显示板插到母板上，如图 2-58 所示。

图 2-58 古诗展示器组装后的开发板

本案例通过分行打印字符串实现文本显示功能。

开发步骤如下:

(1) 在//vendor/hisi/hi3861/hi3861/build/config/usr_config.mk 文件中开启 I^2C。

(2) 在 chapter_02 中创建工程目录 case25-electronic_reader。

(3) 在 case25-electronic_reader 中加载驱动文件 oled_fonts.h、oled_ssd1306.c、oled_ssd1306.h,如图 2-59 所示。

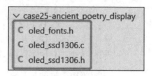

图 2-59 加载驱动文件

(4) 在 case25-electronic_reader 中创建源码文件 electronic_reader_demo.c。

(5) 功能实现,代码如下:

```c
//applications/sample/wifi-iot/app/ohos_50/chapter_02/case25-electronic_reader/electronic_reader_demo.c
#include <stdio.h>
#include <unistd.h>
#include "ohos_init.h"
#include "oled_ssd1306.h"
#include "wifiiot_gpio.h"
#include "wifiiot_gpio_ex.h"
//主功能函数,实现循环打印古诗《春晓》
void entry(void)
{
    //初始化
    OledInit();
    //清屏
    OledFillScreen(0x00);
    //在左上角位置显示字符串 OpenHarmony 50 cases
    OledShowString(0, 0, "OpenHarmony 50 cases", 1);
    sleep(1);
    //古诗名称
    char *title = "Spring Morn";
    //古诗内容
    char *text[] = {
        "Spring slumbers unaware of morn,",
        "All around one hears the bird's call.",
        "Last night, the sound of wind and rain,",
        "How many fallen flowers accounted for? "
    };
    //循环显示
    while (1)
    {
        //清屏
        OledFillScreen(0x00);
```

```
        //在(0,y)位置显示字符串
        OledShowStringSlow(0, 0, title, 2);
        for (int y = 0; y < 8; y++)
        {
            //从 index 为 3 的行(4 行)清屏
            OledFillScreenY(3,0);
            //在(0,y)位置显示字符串
            OledShowStringSlow(0, 3, text[y], 1);
            sleep(1);
        }
        sleep(2);
    }
}
APP_FEATURE_INIT(entry);
```

(6) 在 case25-electronic_reader 中创建模块,构建脚本 BUILD.gn 并初始化模块,代码如下:

```
#//applications/sample/wifi-iot/app/ohos_50/chapter_02/case25-electronic_reader/BUILD.gn
static_library("ch_02_electronic_reader") {
    sources = [
        "electronic_reader_demo.c",
        "oled_ssd1306.c",
    ]
    include_dirs = [
        "//base/iot_hardware/interfaces/kits/wifiiot_lite",
    ]
}
```

(7) 将模块 ch_02_electronic_reader 配置到应用子系统,代码如下:

```
#//applications/sample/wifi-iot/app/BUILD.gn
import("//build/lite/config/component/lite_component.gni")

lite_component("app") {
    features = [
        "ohos_50/chapter_02/case25-electronic_reader:ch_02_electronic_reader",
    ]
}
```

（8）测试：将 LCD 显示板插到母板上，将主控板插到母板上，编译应用模块，将固件烧写到开发板，复位开发板，观察 OLED 显示屏上打印的古诗，如图 2-60 所示。

图 2-60　古诗展示器的运行效果

第3章 OpenHarmony 轻量系统系统开发

本章通过6个案例详细讲解 OpenHarmony 轻量系统中任务、互斥锁、软件定时器、中断处理、内存的开发技术。

3.1 任务

任务是竞争系统资源的最小运行单元。任务可以使用或等待 CPU、使用内存空间等系统资源,并独立于其他任务运行。任务模块可以向用户提供多个任务,实现任务之间的切换和通信,帮助用户管理业务程序流程。

(1) 支持多任务,一个任务表示一个线程。

(2) 任务是抢占式调度机制,同时支持时间片轮转调度方式。

(3) 高优先级的任务可打断低优先级任务,低优先级任务必须在高优先级任务阻塞或结束后才可以得到调度。

(4) 由于系统自身任务需要及时调度,建议用户使用的任务优先级范围是[10,30]。应用级任务建议使用低于系统级任务的优先级。

3.1.1 案例26:计时器

本案例为计时器,实现以秒为单位的计时功能。案例通过调用任务 sleep 函数实现秒数的累加,最后格式化输出数据。

开发步骤如下:

(1) 创建本章源码存放目录 chapter_03。

(2) 在 chapter_03 中创建工程目录 case26-timer。

(3) 在 case26-timer 中创建源码文件 timer_demo.c。

(4) 实现计时器功能,代码如下:

//applications/sample/wifi-iot/app/ohos_50/chapter_03/case26-timer/timer_demo.c

```c
#include <stdio.h>
#include <unistd.h>
#include <ohos_init.h>
int i = 0;              //保存时间
//任务回调执行函数
static void entry(void )
{
    while(i<3600)
    {
        //函数声明在头文件 unistd.h 中,实现任务秒级休眠功能
        sleep(1);
        i++;
        printf("\r%02d:%02d", i/60,i%60);
    }
}
//初始化模块入口函数
APP_FEATURE_INIT(entry);
```

(5) 创建并编写模块,构建脚本 BUILD.gn,代码如下:

```
//applications/sample/wifi-iot/app/ohos_50/chapter_03/case26-timer/BUILD.gn
static_library("ch_03_timer") {
    sources = [
        "timer_demo.c",
    ]
}
```

(6) 将应用模块 ch_03_timer 配置到应用子系统,代码如下:

```
//applications/sample/wifi-iot/app/BUILD.gn
import("//build/lite/config/component/lite_component.gni")
lite_component("app") {
    features = [
        "ohos_50/chapter_03/case26-timer:ch_03_timer",
    ]
}
```

(7) 测试:编译应用模块,将固件烧写到开发板,运行 IPOP 终端工具,以便与开发板相连,复位开发板。观察 IPOP 终端工具的打印信息,如图 3-1 所示。

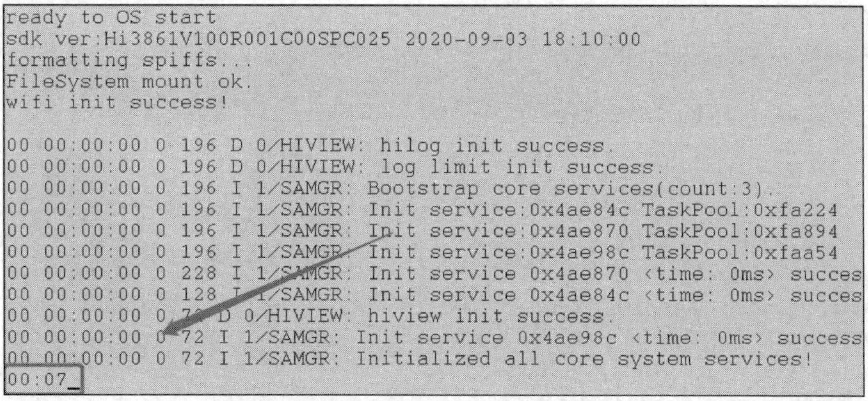

```
ready to OS start
sdk ver:Hi3861V100R001C00SPC025 2020-09-03 18:10:00
formatting spiffs...
FileSystem mount ok.
wifi init success!
00 00:00:00 0 196 D 0/HIVIEW: hilog init success.
00 00:00:00 0 196 D 0/HIVIEW: log limit init success.
00 00:00:00 0 196 I 1/SAMGR: Bootstrap core services(count:3).
00 00:00:00 0 196 I 1/SAMGR: Init service:0x4ae84c TaskPool:0xfa224
00 00:00:00 0 196 I 1/SAMGR: Init service:0x4ae870 TaskPool:0xfa894
00 00:00:00 0 196 I 1/SAMGR: Init service:0x4ae98c TaskPool:0xfaa54
00 00:00:00 0 228 I 1/SAMGR: Init service 0x4ae870 <time: 0ms> succes
00 00:00:00 0 128 I 1/SAMGR: Init service 0x4ae84c <time: 0ms> succes
00 00:00:00 0 72 D 0/HIVIEW: hiview init success.
00 00:00:00 0 72 I 1/SAMGR: Init service 0x4ae98c <time: 0ms> success
00 00:00:00 0 72 I 1/SAMGR: Initialized all core system services!
00:07_
```

图 3-1 计时器的运行效果

3.1.2 案例 27：自动售票系统 V1.0

本案例为自动售票系统的 1.0 版，实现多个任务操作共享数据的功能，即多个终端同时具有售票功能。

案例通过函数 osThreadNew 创建了 3 个任务，模拟 3 个售票终端同时售卖同一种票。开发步骤如下：

（1）在 chapter_03 中创建工程目录 case27-atm_v1.0。

（2）在 case27-atm_v1.0 中创建源码文件 atm_v1.0_demo.c。

（3）实现自动售票机交替售票功能，代码如下：

```c
//applications/sample/wifi-iot/app/ohos_50/chapter_03/case27-atm_v1.0/atm_v1.0_demo.c
#include <stdio.h>
#include <unistd.h>
#include <ohos_init.h>
#include "cmsis_os2.h"
int i = 20;                          //剩余票数
//任务回调执行函数
static void task_function(void *task_name)
{
    while (i > 0)
    {
        //函数声明在头文件 unistd.h 中,实现任务微秒级休眠功能
        usleep(20);
        //售票成功,剩余票数减 1
        i--;
        printf("%s Issue 1 ticket,remaining votes:%d!\r\n", (char *)task_name, i);
    }
}
```

```c
//创建并运行自动售票机
void task_create(char * task_name, osThreadFunc_t func)
{
    //创建任务结构体 attr,配置任务属性
    osThreadAttr_t attr;
    //配置任务名称
    attr.name = task_name;
    attr.attr_bits = 0U;
    attr.cb_mem = NULL;
    attr.cb_size = 0U;
    attr.stack_mem = NULL;
    attr.stack_size = 1024; //将栈配置为1KB
    attr.priority = osPriorityNormal;
    //函数 osThreadNew 声明在头文件 cmsis_os2.h 中,实现任务的创建和运行功能
    if (osThreadNew(func, (void * )task_name, &attr) == NULL)
    {
        printf("%s create failed!\n", task_name);
    }
}
void entry(void)
{
    //函数声明在头文件 unistd.h 中,实现任务秒级休眠功能
    sleep(1);
    //创建并运行3个自动售票机(ATM)
    task_create("ATM_1", task_function);
    task_create("ATM_2", task_function);
    task_create("ATM_3", task_function);
}
//初始化模块入口函数
APP_FEATURE_INIT(entry);
```

(4) 创建并编写模块,构建脚本 BUILD.gn,代码如下:

```
//applications/sample/wifi-iot/app/ohos_50/chapter_02/case27-atm_v1.0/BUILD.gn
static_library("ch_03_atm_v1.0") {
    sources = [
        "atm_v1.0_demo.c",
    ]
    #配置头文件 cmsis_os2.h 的路径
    include_dirs = [
        "//kernel/liteos_m/components/cmsis/2.0",
    ]
}
```

(5) 将应用模块配置到应用子系统,代码如下:

```
//applications/sample/wifi-iot/app/BUILD.gn
import("//build/lite/config/component/lite_component.gni")

lite_component("app") {
    features = [
        "ohos_50/chapter_03/case27-atm_v1.0:ch_03_atm_v1.0",
    ]
}
```

(6) 测试：编译应用模块，将固件烧写到开发板，运行 IPOP 终端工具，以便与开发板相连，复位开发板。观察 IPOP 终端工具的打印信息，如图 3-2 所示。

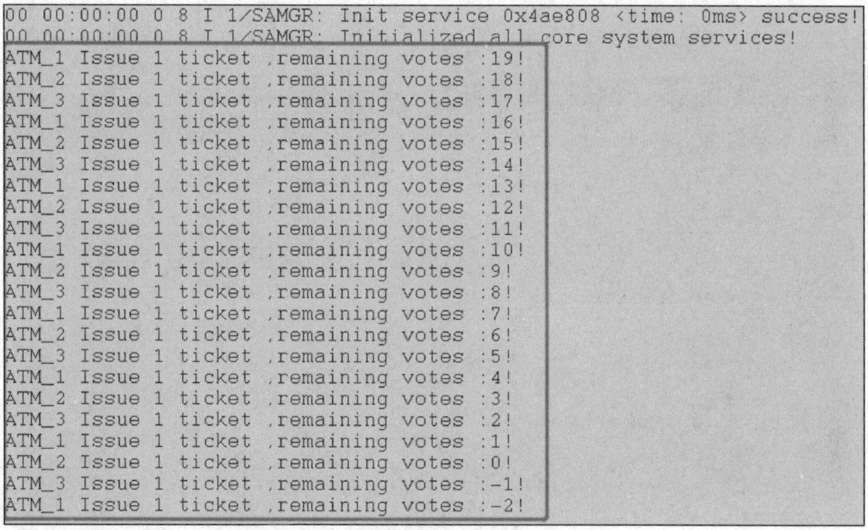

```
00 00:00:00 0 8 I 1/SAMGR: Init service 0x4ae808 <time: 0ms> success!
00 00:00:00 0 8 I 1/SAMGR: Initialized all core system services!
ATM_1 Issue 1 ticket ,remaining votes :19!
ATM_2 Issue 1 ticket ,remaining votes :18!
ATM_3 Issue 1 ticket ,remaining votes :17!
ATM_1 Issue 1 ticket ,remaining votes :16!
ATM_2 Issue 1 ticket ,remaining votes :15!
ATM_3 Issue 1 ticket ,remaining votes :14!
ATM_1 Issue 1 ticket ,remaining votes :13!
ATM_2 Issue 1 ticket ,remaining votes :12!
ATM_3 Issue 1 ticket ,remaining votes :11!
ATM_1 Issue 1 ticket ,remaining votes :10!
ATM_2 Issue 1 ticket ,remaining votes :9!
ATM_3 Issue 1 ticket ,remaining votes :8!
ATM_1 Issue 1 ticket ,remaining votes :7!
ATM_2 Issue 1 ticket ,remaining votes :6!
ATM_3 Issue 1 ticket ,remaining votes :5!
ATM_1 Issue 1 ticket ,remaining votes :4!
ATM_2 Issue 1 ticket ,remaining votes :3!
ATM_3 Issue 1 ticket ,remaining votes :2!
ATM_1 Issue 1 ticket ,remaining votes :1!
ATM_2 Issue 1 ticket ,remaining votes :0!
ATM_3 Issue 1 ticket ,remaining votes :-1!
ATM_1 Issue 1 ticket ,remaining votes :-2!
```

图 3-2　自动售票机 V1.0 的运行效果

3.2　案例 28：自动售票系统 V2.0

本案例是案例 27 的改进版，在案例 27 的基础上使用了互斥锁，以便保证数据被正确操作。涉及的函数为 osMutexAcquire 和 osMutexRelease，在保存数据时使用函数 osMutexAcquire 进行上锁以保证只有当前任务可以操作该数据，操作完成后使用函数 osMutexRelease 释放锁，以便其他任务可以对该数据进行操作。

互斥锁用来保证共享数据操作的完整性。它可以保证共享数据在任一时刻，只能有一个线程执行操作，其他线程处于等待状态。

开发步骤如下：

(1) 在本章源码存放目录 chapter_03 中创建应用模块工程目录 case28-atm_v2.0。

(2) 在应用模块工程目录 case28-atm_v2.0 中创建应用模块源码文件 atm_v2.0_

demo.c。

(3) 实现自动售票机交替售票功能,代码如下:

```c
//applications/sample/wifi-iot/app/ohos_50/chapter_03/case28-atm_v2.0/atm_v2.0_demo.c
#include <stdio.h>
#include <unistd.h>
#include <ohos_init.h>
#include "cmsis_os2.h"
osMutexId_t mutex_id;                   //互斥锁 ID
int i = 20;                             //剩余票数
//任务回调执行函数
static void example_mutex_entry(void * task_name)
{
    while (1)
    {
        //函数 osMutexAcquire 声明在头文件 cmsis_os2.h 中,实现获取互斥锁,锁住其他任务功能
        osMutexAcquire(mutex_id, osWaitForever);
        usleep(20);
        if (i <= 0)
        {
            //函数 osMutexRelease 声明在头文件 cmsis_os2.h 中,实现释放互斥锁功能
            osMutexRelease(mutex_id);
            break;
        }

        //售票成功,剩余票数减 1
        i--;
        printf("%s Issue 1 ticket ,remaining votes : %d!\r\n", (char *)task_name, i);
        //函数 osMutexRelease 声明在头文件 cmsis_os2.h 中,实现释放互斥锁功能
        osMutexRelease(mutex_id);
    }
}
//创建并运行自动售票机
void example_ATM_create(char * task_nane, osThreadFunc_t func)
{
    //创建任务结构体 attr,配置任务属性
    osThreadAttr_t attr;
    //配置任务名称
    attr.name = task_nane;
    attr.attr_bits = 0U;
    attr.cb_mem = NULL;
    attr.cb_size = 0U;
    attr.stack_mem = NULL;
    attr.stack_size = 1024;             //将栈配置为 1KB
    attr.priority = osPriorityNormal;
```

```c
    //创建并运行任务
    if (osThreadNew(func, (void *)task_nane, &attr) == NULL)
    {
        printf("%s create failed!\n", task_nane);
    }
}
void entry(void)
{
    sleep(1);
    //创建并运行3个自动售票机(ATM)
    example_ATM_create("ATM_1", example_mutex_entry);
    example_ATM_create("ATM_2", example_mutex_entry);
    example_ATM_create("ATM_3", example_mutex_entry);
}
//初始化模块入口函数
APP_FEATURE_INIT(entry);
```

(4) 创建并编写模块,构建脚本 BUILD.gn,代码如下:

```
\//applications/sample/wifi-iot/app/ohos_50/chapter_03/case28-atm_v2.0/BUILD.gn
static_library("ch_03_atm_v2.0") {
    sources = [
        "atm_v2.0_demo.c",
    ]
    include_dirs = [
        "//kernel/liteos_m/components/cmsis/2.0",
    ]
}
```

(5) 将应用模块配置到应用子系统,代码如下:

```
\//applications/sample/wifi-iot/app/BUILD.gn
import("//build/lite/config/component/lite_component.gni")
lite_component("app") {
    features = [
        "ohos_50/chapter_03/case28-atm_v2.0:ch_03_atm_v2.0",
    ]
}
```

(6) 测试:编译应用模块,将固件烧写到开发板,运行 IPOP 终端工具,以便与开发板相连,复位开发板。观察 IPOP 终端工具的打印信息,如图 3-3 所示。

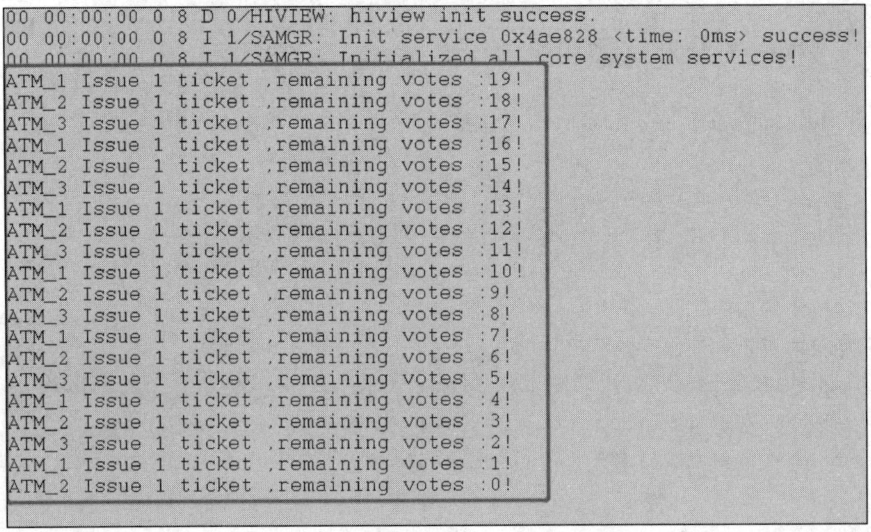

```
00 00:00:00 0 8 D 0/HIVIEW: hiview init success.
00 00:00:00 0 8 I 1/SAMGR: Init service 0x4ae828 <time: 0ms> success!
00 00:00:00 0 8 I 1/SAMGR: Initialized all core system services!
ATM_1 Issue 1 ticket ,remaining votes :19!
ATM_2 Issue 1 ticket ,remaining votes :18!
ATM_3 Issue 1 ticket ,remaining votes :17!
ATM_1 Issue 1 ticket ,remaining votes :16!
ATM_2 Issue 1 ticket ,remaining votes :15!
ATM_3 Issue 1 ticket ,remaining votes :14!
ATM_1 Issue 1 ticket ,remaining votes :13!
ATM_2 Issue 1 ticket ,remaining votes :12!
ATM_3 Issue 1 ticket ,remaining votes :11!
ATM_1 Issue 1 ticket ,remaining votes :10!
ATM_2 Issue 1 ticket ,remaining votes :9!
ATM_3 Issue 1 ticket ,remaining votes :8!
ATM_1 Issue 1 ticket ,remaining votes :7!
ATM_2 Issue 1 ticket ,remaining votes :6!
ATM_3 Issue 1 ticket ,remaining votes :5!
ATM_1 Issue 1 ticket ,remaining votes :4!
ATM_2 Issue 1 ticket ,remaining votes :3!
ATM_3 Issue 1 ticket ,remaining votes :2!
ATM_1 Issue 1 ticket ,remaining votes :1!
ATM_2 Issue 1 ticket ,remaining votes :0!
```

图 3-3　自动售票机 V2.0 的运行效果

3.3　案例 29：软件定时器

本案例为软件定时器，通过函数 osTimerNew 和 osTimerStart 实现软件定时执行指定功能函数的功能。

软件定时器是基于系统 Tick 时钟中断且由软件来模拟的定时器，当经过设定的 Tick 时钟计数值后会触发用户定义的回调函数。定时精度与系统 Tick 时钟的周期有关。

（1）硬件定时器受硬件的限制，数量上不足以满足用户的实际需求，因此为了满足用户需求，需要提供更多的定时器，系统提供软件定时器功能。

（2）软件定时器扩展了定时器的数量，允许创建更多的定时业务。

软件定时器支持的功能如下：

（1）软件定时器创建。

（2）软件定时器启动。

（3）软件定时器停止。

（4）软件定时器删除。

（5）运作机制：

① 软件定时器使用了系统的一个队列和一个任务资源，先进先出。定时时间短的定时器总是比定时时间长的靠近队列头，满足优先被触发的准则。

② 当 Tick 中断到来时，在 Tick 中断处理函数中扫描软件定时器的计时任务，查看是否有定时器超时，如果有，则将超时的定时器记录下来。

③ Tick 中断处理函数结束后，软件定时器任务（优先级为高）被唤醒，在该任务中调用

之前记录下来的定时器的超时回调函数。

(6) 软件定时器提供了两类定时器机制。

① 单次触发定时器：在启动后只会触发一次定时器事件，然后定时器自动删除。

② 周期触发定时器：会周期性地触发定时器事件，直到用户手动地停止定时器，否则将永远持续执行。

开发步骤如下：

(1) 在本章目录 chapter_03 中创建应用模块工程目录 case29-software_timer。

(2) 在应用模块工程目录 case29-software_timer 中创建应用模块源码文件 software_timer_demo.c。

(3) 实现软件定时器功能，代码如下：

```c
//applications/sample/wifi-iot/app/ohos_50/chapter_03/case29-software_timer/software_timer_demo.c
#include "stdio.h"
#include "ohos_init.h"
#include "cmsis_os2.h"
#include <unistd.h>
int g_timercount1 = 0;              //记录定时器1回调函数被执行的次数
int g_timercount2 = 0;              //记录定时器2回调函数被执行的次数
//定时器1任务执行回调函数
void timer1_callback(void * arg)
{
    //函数参数必须使用,否则编译器会报变量未使用错误
    arg = arg;
    g_timercount1++;
    printf("g_timercount1 = %d\n", g_timercount1);
}
//定时器2任务执行回调函数
void timer2_callback(void * arg)
{
    arg = arg;
    g_timercount2++;
    printf("g_timercount2 = %d\n", g_timercount2);
}
void entry(void)
{
    sleep(1);
    //创建定时器id数组
    osTimerId_t timer_id[2];
    //创建定时器回调函数数组
    osTimerFunc_t timer_callback[2] = {timer1_callback,timer2_callback};
    //创建定时器属性结构体数组
    osTimerAttr_t timer_att[2] = {{"timer1",0,NULL,0},{"timer2",0,NULL,0}};
```

```c
        //创建定时器回调间隔的ticks数组,1tick的时长由芯片主频决定,如Hi3861芯片中1tick约
        //等于10ms
        unsigned int ticks[2] = {300,500};
        for(int i = 0;i < 2;i++){
            //函数osTimerNew声明在头文件cmsis_os2.h中,实现创建定时器功能
            timer_id[i] = osTimerNew(timer_callback[i],osTimerPeriodic,NULL,timer_att + i);

            //函数osTimerStart声明在头文件cmsis_os2.h中,实现启用定时器功能
            osTimerStart(timer_id[i],ticks[i]);
        }

        sleep(60);
        //判断定时器是否运行,如果运行,则停止并删除
        for(int i = 0;i < 2;i++){
            //函数osTimerIsRunning声明在头文件cmsis_os2.h中,用于检查定时器是否运行
            if(osTimerIsRunning(timer_id[i])){
                //函数osTimerStop声明在头文件cmsis_os2.h中,实现停止定时器功能
                osTimerStop(timer_id[i]);
                //函数osTimerDelete声明在头文件cmsis_os2.h中,实现删除定时器功能
                osTimerDelete(timer_id[i]);
            }
        }
}
//创建并运行定时器任务
void example_timer_create(void )
{
    //创建osThreadAttr_t结构体attr,用于存放任务参数
    osThreadAttr_t attr;
    //设置任务名称
    attr.name = "timer";
    attr.attr_bits = 0U;
    attr.cb_mem = NULL;
    attr.cb_size = 0U;
    attr.stack_mem = NULL;
    attr.stack_size = 1024; //将栈配置为1KB
    attr.priority = osPriorityNormal;
    //创建并运行任务
    if (osThreadNew((osThreadFunc_t)entry, NULL, &attr) == NULL)
    {
        printf("timer task create failed!\n");
    }
}
//初始化模块入口函数
APP_FEATURE_INIT(example_timer_create);
```

(4）创建并编写模块,构建脚本 BUILD.gn,代码如下：

```
//applications/sample/wifi-iot/app/ohos_50/chapter_03/case29-software_timer/BUILD.gn
static_library("ch_03_software_timer") {
    sources = [
        "software_timer_demo.c",
    ]
    include_dirs = [
        "//kernel/liteos_m/components/cmsis/2.0",
    ]
}
```

(5）将应用模块配置到应用子系统,代码如下：

```
//applications/sample/wifi-iot/app/BUILD.gn
import("//build/lite/config/component/lite_component.gni")
lite_component("app") {
    features = [
        "ohos_50/chapter_03/case29-software_timer:ch_03_software_timer",
    ]
}
```

(6）测试：编译应用模块,将固件烧写到开发板,运行 IPOP 终端工具,以便与开发板相连,复位开发板。观察 IPOP 终端工具的打印信息,如图 3-4 所示。

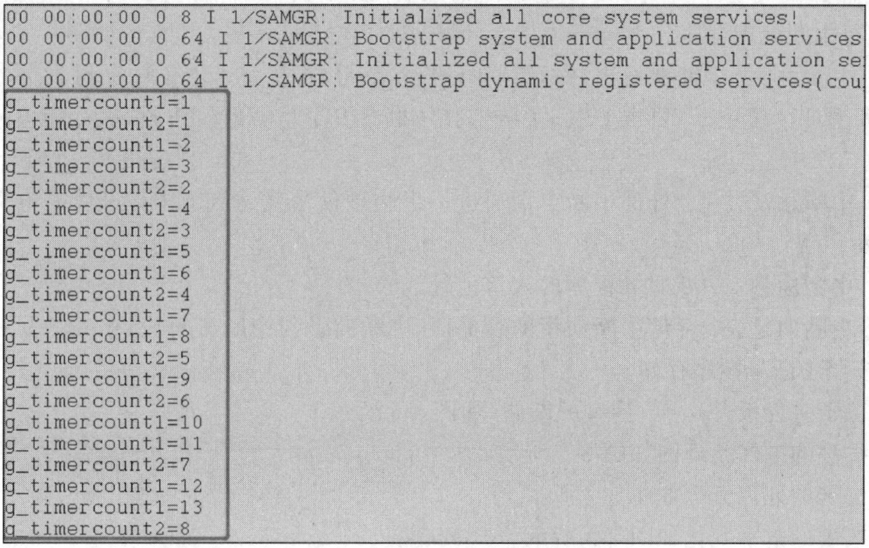

```
00 00:00:00 0 8 I 1/SAMGR: Initialized all core system services!
00 00:00:00 0 64 I 1/SAMGR: Bootstrap system and application services
00 00:00:00 0 64 I 1/SAMGR: Initialized all system and application se
00 00:00:00 0 64 I 1/SAMGR: Bootstrap dynamic registered services(cou
g_timercount1=1
g_timercount2=1
g_timercount1=2
g_timercount1=3
g_timercount2=2
g_timercount1=4
g_timercount2=3
g_timercount1=5
g_timercount1=6
g_timercount2=4
g_timercount1=7
g_timercount1=8
g_timercount2=5
g_timercount1=9
g_timercount2=6
g_timercount1=10
g_timercount1=11
g_timercount2=7
g_timercount1=12
g_timercount1=13
g_timercount2=8
```

图 3-4　软件定时器的运行效果

3.4 案例30：按键中断处理

本案例为按键中断处理，通过函数 GpioRegisterIsrFunc 和 GpioUnregisterIsrFunc 实现按键 S2 中断注册与注销功能。

中断是指 CPU 暂停执行当前程序，转而执行新程序的过程。与中断相关的硬件可以划分为 3 类。

（1）设备：发起中断的源，当设备需要请求 CPU 时，产生一个中断信号，该信号连接至中断控制器。

（2）中断控制器：接收中断输入并上报给 CPU。可以设置中断源的优先级、触发方式、打开和关闭等操作。

（3）CPU：判断和执行中断任务。

与中断相关的名词解释如下。

（1）中断号：每个中断请求信号都会有特定的标志，使计算机能够判断是哪个设备提出的中断请求，这个标志就是中断号。

（2）中断请求："紧急事件"需向 CPU 提出申请（发一个电脉冲信号），要求中断，以及要求 CPU 暂停当前执行的任务，转而处理该"紧急事件"，这一申请过程称为中断申请。

（3）中断优先级：为使系统能够及时响应并处理所有中断，系统根据中断事件的重要性和紧迫程度，将中断源分为若干个级别，称作中断优先级。系统中所有的中断源优先级相同，不支持中断嵌套或抢占。

（4）中断处理程序：当外设产生中断请求后，CPU 暂停当前的任务，转而响应中断申请，即执行中断处理程序。

（5）中断触发：中断源发出并送给 CPU 控制信号，将接口卡上的中断触发器置"1"，表明该中断源产生了中断，要求 CPU 去响应该中断，CPU 暂停当前任务，执行相应的中断处理程序。

（6）中断触发类型：外部中断申请通过一个物理信号发送到 CPU，可以是电平触发或边沿触发。

（7）中断向量：中断服务程序的入口地址。

（8）中断向量表：存储中断向量的存储区，中断向量与中断号对应，中断向量在中断向量表中按照中断号顺序存储。

S2 位于主控板 Type-C 接口的左侧，如图 3-5 所示。

S2 按键的电路原理图如图 3-6 所示。

S2 按键的处理流程如图 3-7 所示。

S2 按键的电平变化如图 3-8 所示。

开发步骤如下：

（1）在本章目录 chapter_03 中创建应用模块工程目录 case30-isr。

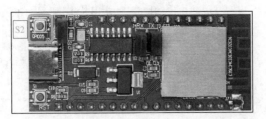

图 3-5　S2 按键

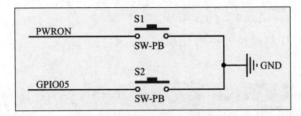

图 3-6　S2 按键的电路原理图

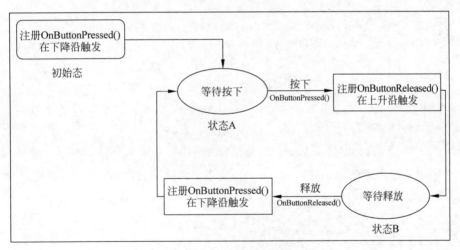

图 3-7　S2 按键的处理流程

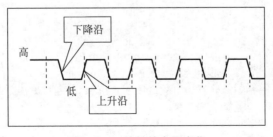

图 3-8　S2 按键的电平变化

(2) 在应用模块工程目录 case30-isr 中创建应用模块源码文件 isr.c。
(3) 引用必要的头文件,代码如下:

```c
//applications/sample/wifi-iot/app/ohos_50/chapter_02/case30-isr/isr.c
#include <stdio.h>
#include <unistd.h>
#include <ohos_init.h>
#include "cmsis_os2.h"
#include "wifiiot_gpio.h"
#include "wifiiot_gpio_ex.h"
```

(4) 在 isr.c 文件中创建按键中断处理回调函数 on_button_pressed,实现中断处理功能,代码如下:

```c
//applications/sample/wifi-iot/app/ohos_50/chapter_02/case30-isr/isr.c
//按键中断处理回调函数
static void on_button_pressed(char *arg)
{
    (void)arg;
    //函数声明在头文件 wifiiot_gpio.h 中,实现给 GPIO5 引脚注销中断功能
    GpioUnregisterIsrFunc(WIFI_IOT_IO_NAME_GPIO_5);
    printf("key is pressed\r\n");
    sleep(1);
    //函数声明在头文件 wifiiot_gpio.h 中,实现给 GPIO5 引脚注册下降沿触发中断
    GpioRegisterIsrFunc(WIFI_IOT_IO_NAME_GPIO_5, WIFI_IOT_INT_TYPE_EDGE, WIFI_IOT_GPIO_EDGE_FALL_LEVEL_LOW, on_button_pressed, NULL);
}
```

(5) 创建主任务函数 example_isr_entry,实现 GPIO 引脚的初始化、功能设置、IO 方向设置、注册中断功能,代码如下:

```c
//applications/sample/wifi-iot/app/ohos_50/chapter_02/case30-isr/isr.c
static void example_isr_entry(void)
{
    //函数声明在头文件 wifiiot_gpio.h 中,实现初始化 GPIO 引脚功能
    GpioInit();
    //函数声明在头文件 wifiiot_gpio.h 中,实现给 GPIO5 引脚设置 GPIO 功能
    IoSetFunc(WIFI_IOT_IO_NAME_GPIO_5, WIFI_IOT_IO_FUNC_GPIO_5_GPIO);
    //函数声明在头文件 wifiiot_gpio.h 中,实现将 GPIO5 引脚 IO 方向设置为输入
    GpioSetDir(WIFI_IOT_IO_NAME_GPIO_5, WIFI_IOT_GPIO_DIR_IN);
    //函数声明在头文件 wifiiot_gpio_ex.h 中,实现给 GPIO5 引脚默认为 3.3V 高电平
    IoSetPull(WIFI_IOT_IO_NAME_GPIO_5, WIFI_IOT_IO_PULL_UP);
    //函数声明在头文件 wifiiot_gpio.h 中,实现给 GPIO5 引脚注册下降沿触发中断
    GpioRegisterIsrFunc(WIFI_IOT_IO_NAME_GPIO_5, WIFI_IOT_INT_TYPE_EDGE, WIFI_IOT_GPIO_EDGE_FALL_LEVEL_LOW, on_button_pressed, NULL);
```

```
        //主任务处于等待状态
    while (1)
    {
        usleep(1000);
    }
}
```

(6) 创建函数 example_isr_create,在函数中创建任务,将任务执行函数设置为 example_isr_entry,将模块入口函数初始化为 example_isr_create,代码如下:

```
//applications/sample/wifi-iot/app/ohos_50/chapter_02/case30-isr/isr.c
static void example_isr_create(void)
{
    osThreadAttr_t attr;
    attr.name = "isr_task";
    attr.attr_bits = 0U;
    attr.cb_mem = NULL;
    attr.cb_size = 0U;
    attr.stack_mem = NULL;
    attr.stack_size = 1024;
    attr.priority = osPriorityNormal;
    if (osThreadNew((osThreadFunc_t)example_isr_entry, NULL, &attr) == NULL)
    {
        printf("[example_isr_create] Failed to create isr_task!\n");
    }
}

APP_FEATURE_INIT(example_isr_create);
```

(7) 创建并编写模块,构建脚本 BUILD.gn,代码如下:

```
//applications/sample/wifi-iot/app/ohos_50/chapter_02/case30-isr/BUILD.gn
static_library("ch_03_isr") {
    sources = [
        "isr.c",
    ]
    include_dirs = [
        "//kernel/liteos_m/components/cmsis/2.0",
        "//base/iot_hardware/interfaces/kits/wifiiot_lite",
    ]
}
```

(8) 将应用模块配置到应用子系统,代码如下:

```
//applications/sample/wifi-iot/app/BUILD.gn
```

```
import("//build/lite/config/component/lite_component.gni")
lite_component("app") {
    features = [
        "ohos_50/chapter_02/case30-isr:ch_03_isr",
    ]
}
```

（9）测试：编译应用模块，将固件烧写到开发板，运行 IPOP 终端工具，以便与开发板相连，复位开发板，然后多次按 S2 按键。观察 IPOP 终端工具的打印信息，如图 3-9 所示。

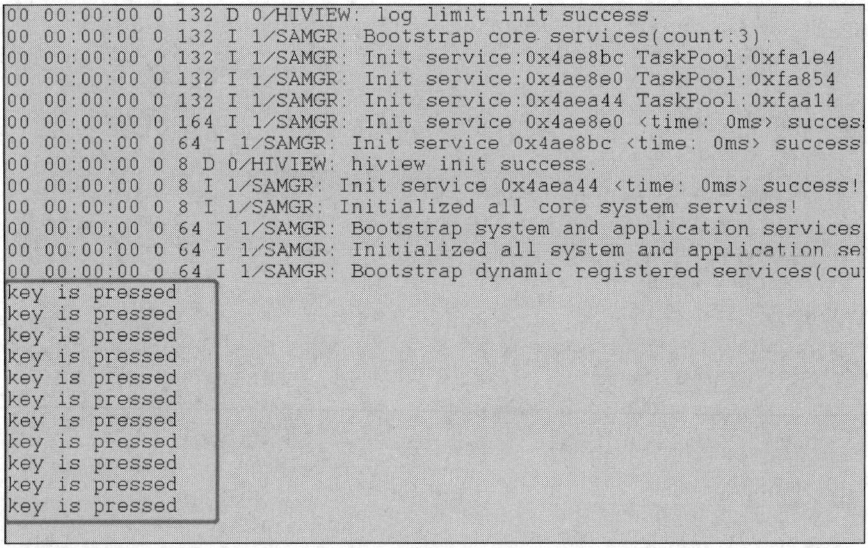

```
00 00:00:00 0 132 D 0/HIVIEW: log limit init success.
00 00:00:00 0 132 I 1/SAMGR: Bootstrap core services(count:3).
00 00:00:00 0 132 I 1/SAMGR: Init service:0x4ae8bc TaskPool:0xfa1e4
00 00:00:00 0 132 I 1/SAMGR: Init service:0x4ae8e0 TaskPool:0xfa854
00 00:00:00 0 132 I 1/SAMGR: Init service:0x4aea44 TaskPool:0xfaa14
00 00:00:00 0 164 I 1/SAMGR: Init service 0x4ae8e0 <time: 0ms> succes
00 00:00:00 0  64 I 1/SAMGR: Init service 0x4ae8bc <time: 0ms> success
00 00:00:00 0   8 D 0/HIVIEW: hiview init success.
00 00:00:00 0   8 I 1/SAMGR: Init service 0x4aea44 <time: 0ms> success!
00 00:00:00 0   8 I 1/SAMGR: Initialized all core system services!
00 00:00:00 0  64 I 1/SAMGR: Bootstrap system and application services
00 00:00:00 0  64 I 1/SAMGR: Initialized all system and application se
00 00:00:00 0  64 I 1/SAMGR: Bootstrap dynamic registered services(cou
key is pressed
key is pressed
key is pressed
key is pressed
key is pressed
key is pressed
key is pressed
key is pressed
key is pressed
key is pressed
```

图 3-9　按键中断的处理效果

3.5　案例 31：内存申请与释放

本案例为内存申请与释放，实现动态内存申请、操作和释放功能。由于 OpenHarmony 轻量系统支持标准 C 库，所以本案例讲解使用标准 C 库中的函数 malloc 和 free 实现内存的申请与释放功能。

内存管理模块管理系统的内存资源，通过对内存的申请/释放操作来管理用户和 OS 对内存的使用，使内存的利用率和效率最优，最大限度地解决系统的内存碎片问题，其中，OS 的内存管理为动态内存管理，提供内存初始化、分配、释放等功能。动态内存是指在动态内存池中分配用户指定大小的内存块。

（1）优点：按需分配。

（2）缺点：内存池中可能出现碎片。

开发步骤如下：

(1) 在本章目录 chapter_03 中创建应用模块工程目录 case31-memory。
(2) 在应用模块工程目录 case31-memory 中创建应用模块源码文件 memory.c。
(3) 引用必要的头文件,代码如下:

```c
//applications/sample/wifi-iot/app/ohos_50/chapter_03/case31-memory/memory.c
#include <stdio.h>
#include <stdlib.h>
#include <string.h>
#include <unistd.h>
#include <ohos_init.h>
#include "cmsis_os2.h"
```

(4) 在 memory.c 文件中创建函数 example_mem_entry,调用函数 malloc 和 free 实现内存的申请与释放功能,代码如下:

```c
//applications/sample/wifi-iot/app/ohos_50/chapter_03/case31-memory/memory.c
static void example_mem_entry(void)
{
    //申请100字节的内存空间
    void * ptr = malloc(100);
    if(ptr == NULL ){
        printf("Malloc failed!\n\r");
        return;
    }
    //以0进行内存初始化
    memset(ptr,0,100);
    //将字符串复制到内存
    strcpy(ptr,"hello world!\r\n");
    //打印内存内容
    printf("ptr:%s\r\n",ptr);
    //释放内存,防止内存泄漏
    free(ptr);
    ptr = NULL;
}
```

(5) 创建函数 example_mem_create,在函数中创建任务,将任务执行函数设置为 example_mem_entry,将模块入口函数初始化为 example_mem_create,代码如下:

```c
//applications/sample/wifi-iot/app/ohos_50/chapter_03/case31-memory/memory.c
static void example_mem_create(void)
```

```
{
    osThreadAttr_t attr;
    sleep(1);
    attr.name = "memory_task";
    attr.attr_bits = 0U;
    attr.cb_mem = NULL;
    attr.cb_size = 0U;
    attr.stack_mem = NULL;
    attr.stack_size = 1024;
    attr.priority = osPriorityNormal;
    if (osThreadNew((osThreadFunc_t)example_mem_entry, NULL, &attr) == NULL)
    {
        printf("[example_mem_create] Failed to create memory task!\n");
    }
}
APP_FEATURE_INIT(example_mem_create);
```

(6) 创建并编写模块,构建脚本 BUILD.gn,代码如下:

```
//applications/sample/wifi-iot/app/ohos_50/chapter_03/case31-memory/BUILD.gn
static_library("ch_03_mem") {
    sources = [
        "memory.c",
    ]
    include_dirs = [
        "//kernel/liteos_m/components/cmsis/2.0",
    ]
}
```

(7) 将应用模块配置到应用子系统,代码如下:

```
//applications/sample/wifi-iot/app/BUILD.gn
import("//build/lite/config/component/lite_component.gni")
lite_component("app") {
    features = [
        "ohos_50/chapter_03/case31-memory:ch_03_mem",
    ]
}
```

(8) 测试:编译应用模块,将固件烧写到开发板,运行 IPOP 终端工具,以便与开发板相连，复位开发板。观察 IPOP 终端工具的打印信息，如图 3-10 所示。

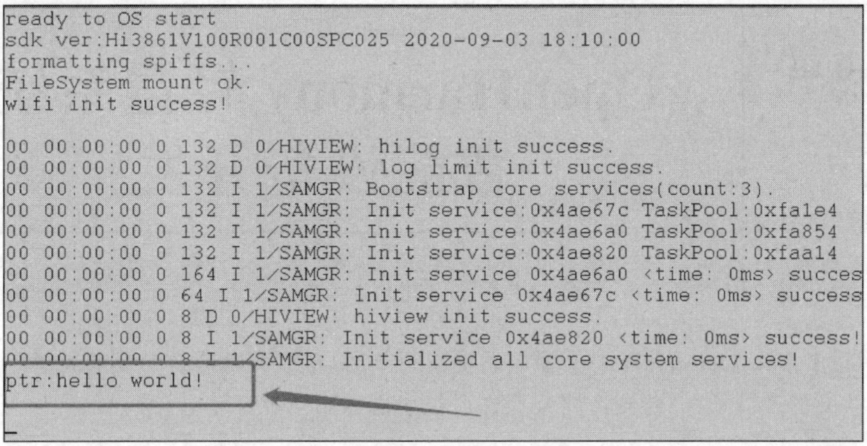

```
ready to OS start
sdk ver:Hi3861V100R001C00SPC025 2020-09-03 18:10:00
formatting spiffs...
FileSystem mount ok.
wifi init success!

00 00:00:00 0 132 D 0/HIVIEW: hilog init success.
00 00:00:00 0 132 D 0/HIVIEW: log limit init success.
00 00:00:00 0 132 I 1/SAMGR: Bootstrap core services(count:3).
00 00:00:00 0 132 I 1/SAMGR: Init service:0x4ae67c TaskPool:0xfa1e4
00 00:00:00 0 132 I 1/SAMGR: Init service:0x4ae6a0 TaskPool:0xfa854
00 00:00:00 0 132 I 1/SAMGR: Init service:0x4ae820 TaskPool:0xfaa14
00 00:00:00 0 164 I 1/SAMGR: Init service 0x4ae6a0 <time: 0ms> succes
00 00:00:00 0 64 I 1/SAMGR: Init service 0x4ae67c <time: 0ms> success
00 00:00:00 0 8 D 0/HIVIEW: hiview init success.
00 00:00:00 0 8 I 1/SAMGR: Init service 0x4ae820 <time: 0ms> success!
00 00:00:00 0 8 I 1/SAMGR: Initialized all core system services!
ptr:hello world!
_
```

图 3-10　内存申请与释放

第 4 章 OpenHarmony 轻量系统智能设备开发

本章讲解 5 个 OpenHarmony 轻量系统应用产品,希望可以给读者带来灵感,开发出更多、更有意义的产品,为完善国产鸿蒙操作系统生态做出贡献。

4.1 案例 32:智能雨刷

本案例为智能雨刷,实现雨水量监测和雨刷速度智能调节功能。核心模块为舵机和雨滴传感器。

将雨滴探测器的 AO、VCC、GND 引脚分别连接到母板的 P12、V33、GND 上,将舵机的黄、红、棕线分别连接到母板的 P02、V33、GND 上。

开发步骤如下:

(1) 创建本章源码存放目录 chapter_04。

(2) 在本章目录 chapter_04 中创建应用模块工程目录 case32-smart_wiper。

(3) 在应用模块工程目录 case32-smart_wiper 中创建应用模块源码文件 smart_wiper_demo.c。

(4) 引用必要的头文并声明必要的全局变量,代码如下:

```c
//applications/sample/wifi-iot/app/ohos_50/chapter_04/case32-smart_wiper/smart_wiper_demo.c
#include <stdio.h>
#include <unistd.h>
#include <ohos_init.h>
#include <wifiiot_adc.h>
#include "cmsis_os2.h"
#include "wifiiot_gpio.h"
#include "wifiiot_gpio_ex.h"

//将使用的 ADC CHANNEL 序号声明为 0
#define ADC_CHANNEL WIFI_IOT_ADC_CHANNEL_0
#define Pin WIFI_IOT_IO_NAME_GPIO_2
```

```c
//用来保存读取到的模拟数据
unsigned short data = 0;
```

(5) 实现感应雨刷功能,代码如下:

```c
//applications/sample/wifi-iot/app/ohos_50/chapter_04/case32-smart_wiper/smart_wiper_demo.c
void init(void)
{
    GpioInit();
    IoSetFunc(Pin, WIFI_IOT_IO_FUNC_GPIO_2_GPIO);
    GpioSetDir(Pin, WIFI_IOT_GPIO_DIR_OUT);
}
//任务回调执行函数
static void main_task(void *argv)
{
    argv = argv;
    init();
    while (1)
    {
        AdcRead(ADC_CHANNEL, &data, WIFI_IOT_ADC_EQU_MODEL_4, WIFI_IOT_ADC_CUR_BAIS_DEFAULT, 0);

        printf("adc_read_data:%d\r\n", data);
        //当data大于1200时表示没有雨水
        if (data > 1200)
        {
            sleep(1);
        }
        else
        {
            sleep(data/200);
            GpioSetOutputVal(Pin, WIFI_IOT_GPIO_VALUE1);
            usleep(2500);
            GpioSetOutputVal(Pin, WIFI_IOT_GPIO_VALUE0);
            usleep(500000);
            GpioSetOutputVal(Pin, WIFI_IOT_GPIO_VALUE1);
            usleep(500);
            GpioSetOutputVal(Pin, WIFI_IOT_GPIO_VALUE0);
        }
        sleep(2);
    }
}
void entry(void)
{
```

```
    //创建任务结构体 attr,配置任务属性
    osThreadAttr_t attr;
    //配置任务名称
    attr.name = "main_task";
    attr.attr_bits = 0U;
    attr.cb_mem = NULL;
    attr.cb_size = 0U;
    attr.stack_mem = NULL;
    attr.stack_size = 1024;              //将栈配置为1KB
    attr.priority = osPriorityNormal;
    //函数 osThreadNew 声明在头文件 cmsis_os2.h 中,实现任务的创建和运行功能
    if (osThreadNew(main_task, NULL, &attr) == NULL)
    {
        printf("%s create failed!\n", "main_task");
    }
}
//初始化模块入口函数
APP_FEATURE_INIT(entry);
```

(6) 创建并编写模块,构建脚本 BUILD.gn,代码如下:

```
//applications/sample/wifi-iot/app/ohos_50/chapter_04/case32-smart_wiper/BUILD.gn
static_library("ch_04_smart_wiper") {
    sources = [
        "smart_wiper_demo.c",
    ]
    include_dirs = [
        "//base/iot_hardware/interfaces/kits/wifiiot_lite",
        "//kernel/liteos_m/components/cmsis/2.0",
    ]
}
```

(7) 将应用模块配置到应用子系统,代码如下:

```
//applications/sample/wifi-iot/app/BUILD.gn
import("//build/lite/config/component/lite_component.gni")
lite_component("app") {
    features = [
        "ohos_50/chapter_04/case32-smart_wiper:ch_04_smart_wiper",
    ]
}
```

(8) 测试:编译应用模块,将固件烧写到开发板,运行 IPOP 终端工具,以便与开发板相连,复位开发板,在雨滴探测模块上滴水。观察 IPOP 终端工具的打印信息,以及舵机转动的角度,如图 4-1 所示。

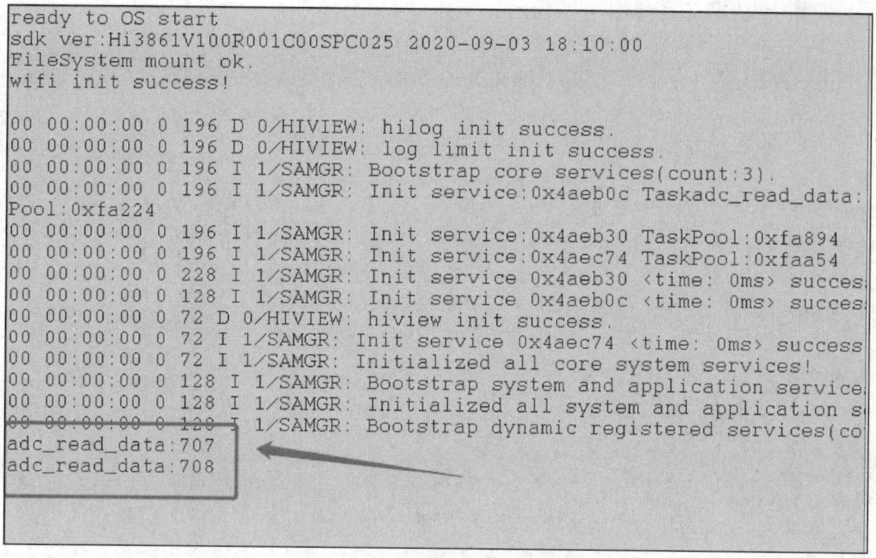

```
ready to OS start
sdk ver:Hi3861V100R001C00SPC025 2020-09-03 18:10:00
FileSystem mount ok.
wifi init success!
00 00:00:00 0 196 D 0/HIVIEW: hilog init success.
00 00:00:00 0 196 D 0/HIVIEW: log limit init success.
00 00:00:00 0 196 I 1/SAMGR: Bootstrap core services(count:3).
00 00:00:00 0 196 I 1/SAMGR: Init service:0x4aeb0c Taskadc_read_data:
Pool:0xfa224
00 00:00:00 0 196 I 1/SAMGR: Init service:0x4aeb30 TaskPool:0xfa894
00 00:00:00 0 196 I 1/SAMGR: Init service 0x4aec74 TaskPool:0xfaa54
00 00:00:00 0 228 I 1/SAMGR: Init service 0x4aeb30 <time: 0ms> succes
00 00:00:00 0 128 I 1/SAMGR: Init service 0x4aeb0c <time: 0ms> succes
00 00:00:00 0 72 D 0/HIVIEW: hiview init success.
00 00:00:00 0 72 I 1/SAMGR: Init service 0x4aec74 <time: 0ms> success
00 00:00:00 0 72 I 1/SAMGR: Initialized all core system services!
00 00:00:00 0 128 I 1/SAMGR: Bootstrap system and application service.
00 00:00:00 0 128 I 1/SAMGR: Initialized all system and application s
00 00:00:00 0 128 I 1/SAMGR: Bootstrap dynamic registered services(co
adc_read_data:707
adc_read_data:708
```

图 4-1 感应雨刷的运行效果

4.2 案例33：智能雷达

本案例为智能雷达，实现雷达智能旋转、探测功能，读者可以在此基础上再加一个舵机，以便实现立体探测功能。核心模块为舵机和超声波探测器。

将超声波探测器的 VCC、Trig、Echo、GND 引脚分别连接到母板的 V33、P07、P08、GND 上，将舵机的黄、红、棕线分别连接到母板的 GPIO02、VCC、GND 上。

开发步骤如下：

（1）在本章目录 chapter_04 中创建应用模块工程目录 case33-radar。
（2）在应用模块工程目录 case33-radar 中创建应用模块源码文件 radar_demo.c。
（3）引用必要的头文并声明必要的全局变量，代码如下：

```
//applications/sample/wifi-iot/app/ohos_50/chapter_04/case33-radar/radar_demo.c
#include <stdio.h>
#include <stdlib.h>
#include <unistd.h>
#include "hi_time.h"
#include "ohos_init.h"
#include "cmsis_os2.h"
#include "wifiiot_gpio.h"
#include "wifiiot_gpio_ex.h"
#define GPIO2 2
#define DELAY 1000
```

(4) 实现雷达功能，代码如下：

```c
//applications/sample/wifi-iot/app/ohos_50/chapter_04/case33-radar/radar_demo.c
void init(void)
{
    GpioInit();
    IoSetFunc(WIFI_IOT_IO_NAME_GPIO_7, WIFI_IOT_IO_FUNC_GPIO_7_GPIO);
    GpioSetDir(WIFI_IOT_IO_NAME_GPIO_7, WIFI_IOT_GPIO_DIR_OUT);
    IoSetFunc(WIFI_IOT_IO_NAME_GPIO_8, WIFI_IOT_IO_FUNC_GPIO_8_GPIO);
    GpioSetDir(WIFI_IOT_IO_NAME_GPIO_8, WIFI_IOT_GPIO_DIR_IN);
    IoSetFunc(WIFI_IOT_IO_NAME_GPIO_2, WIFI_IOT_IO_FUNC_GPIO_2_GPIO);
    GpioSetDir(WIFI_IOT_IO_NAME_GPIO_2, WIFI_IOT_GPIO_DIR_OUT);
}
void get_distance(void)
{
    float distance = 0.0;
    unsigned int flag = 0;
    wifiiotGpioValue value;
    static unsigned long start_time = 0, time = 0;
    //GPIO07输出高电平,超声波传感器开始测距
    GpioSetOutputVal(WIFI_IOT_IO_NAME_GPIO_7, WIFI_IOT_GPIO_VALUE1);
    usleep(20);
    GpioSetOutputVal(WIFI_IOT_IO_NAME_GPIO_7, WIFI_IOT_GPIO_VALUE0);
    while (1)
    {
        GpioGetInputVal(WIFI_IOT_IO_NAME_GPIO_8, &value);
        if (value == WIFI_IOT_GPIO_VALUE1 && flag == 0)
        {
            start_time = hi_get_us();
            flag = 1;
        }
        if (value == WIFI_IOT_GPIO_VALUE0 && flag == 1)
        {
            time = hi_get_us() - start_time;
            start_time = 0;
            break;
        }
    }
    //计算距离
    distance = time * 0.034 / 2;
    printf("\r\ndistance is %.2f cm", distance);
}
void radar_thread(void *parame)
{
    (void)parame;
    printf("The steering gear is centered\r\n");
```

```c
        init();
        while (1)
        {
            for (int i = 1000; i <= 2000; i += 8)
            {
                GpioSetOutputVal(GPIO2, WIFI_IOT_GPIO_VALUE1);
                hi_udelay(i);
                GpioSetOutputVal(GPIO2, WIFI_IOT_GPIO_VALUE0);
                if (i % 100 == 0)
                    get_distance();
                usleep(DELAY);
            }
            for (int i = 2000; i >= 1000; i -= 8)
            {
                GpioSetOutputVal(GPIO2, WIFI_IOT_GPIO_VALUE1);
                hi_udelay(i);
                GpioSetOutputVal(GPIO2, WIFI_IOT_GPIO_VALUE0);
                if (i % 100 == 0)
                    get_distance();
                usleep(DELAY);
            }
        }
}
static void entry(void)
{
    osThreadAttr_t attr;
    attr.name = "radar_thread";
    attr.attr_bits = 0U;
    attr.cb_mem = NULL;
    attr.cb_size = 0U;
    attr.stack_mem = NULL;
    attr.stack_size = 10240;
    attr.priority = osPriorityNormal;
    if (osThreadNew(radar_thread, NULL, &attr) == NULL)
    {
        printf("[radar_thread] Failed to create radar_thread!\n");
    }
}
APP_FEATURE_INIT(entry);
```

(5) 创建并编写模块，构建脚本 BUILD.gn，代码如下：

```
//applications/sample/wifi-iot/app/ohos_50/chapter_04/case33-radar/BUILD.gn
static_library("ch_04_radar") {
    sources = [
```

```
        "radar_demo.c"
    ]
    include_dirs = [
        "//kernel\liteos_m\components\cmsis\2.0",
        "//base/iot_hardware/interfaces/kits/wifiiot_lite",
    ]
}
```

（6）将应用模块配置到应用子系统，代码如下：

```
//applications/sample/wifi-iot/app/BUILD.gn
import("//build/lite/config/component/lite_component.gni")
lite_component("app") {
    features = [
        "ohos_50/chapter_04/case33-radar:ch_04_radar",
    ]
}
```

（7）测试：编译应用模块，将固件烧写到开发板，运行 IPOP 终端工具，以便与开发板相连，复位开发板。观察 IPOP 终端工具的打印信息，以及舵机转动的角度，如图 4-2 所示。

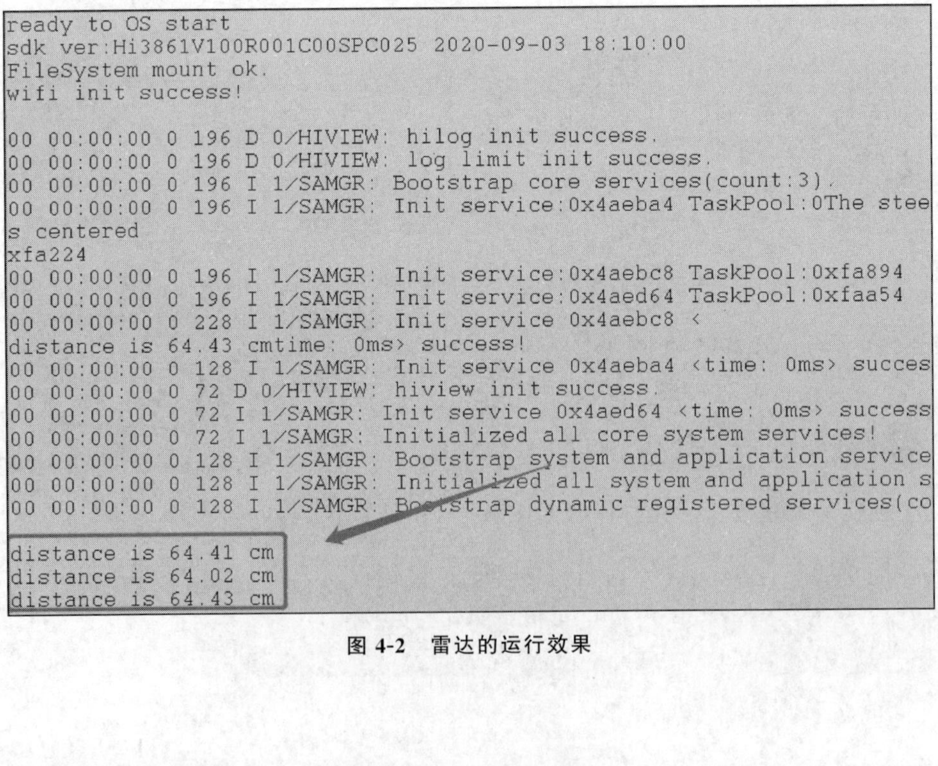

```
ready to OS start
sdk ver:Hi3861V100R001C00SPC025 2020-09-03 18:10:00
FileSystem mount ok.
wifi init success!

00 00:00:00 0 196 D 0/HIVIEW: hilog init success.
00 00:00:00 0 196 D 0/HIVIEW: log limit init success.
00 00:00:00 0 196 I 1/SAMGR: Bootstrap core services(count:3).
00 00:00:00 0 196 I 1/SAMGR: Init service:0x4aeba4 TaskPool:0The stee
s centered
xfa224
00 00:00:00 0 196 I 1/SAMGR: Init service:0x4aebc8 TaskPool:0xfa894
00 00:00:00 0 196 I 1/SAMGR: Init service:0x4aed64 TaskPool:0xfaa54
00 00:00:00 0 228 I 1/SAMGR: Init service 0x4aebc8 <
distance is 64.43 cmtime: 0ms> success!
00 00:00:00 0 128 I 1/SAMGR: Init service 0x4aeba4 <time: 0ms> succes
00 00:00:00 0 72 D 0/HIVIEW: hiview init success.
00 00:00:00 0 72 I 1/SAMGR: Init service 0x4aed64 <time: 0ms> success
00 00:00:00 0 72 I 1/SAMGR: Initialized all core system services!
00 00:00:00 0 128 I 1/SAMGR: Bootstrap system and application service
00 00:00:00 0 128 I 1/SAMGR: Initialized all system and application s
00 00:00:00 0 128 I 1/SAMGR: Bootstrap dynamic registered services(co
distance is 64.41 cm
distance is 64.02 cm
distance is 64.43 cm
```

图 4-2　雷达的运行效果

4.3 案例34：智能人体感应灯

本案例为智能人体感应灯，实现人员监测、智能照明功能。核心模块为人体红外感应器和 LED。

将主控板插到母板上，再将人体红外感应板插到母板上，如图 4-3 所示。

图 4-3　人体感应灯组装后的开发板

开发步骤如下：

（1）在本章目录 chapter_04 中创建应用模块工程目录 case34-induction_lamp。

（2）在应用模块工程目录 case34-induction_lamp 中创建应用模块源码文件 induction_lamp_demo.c。

（3）引用必要的头文件，代码如下：

```
//applications/sample/wifi-iot/app/ohos_50/chapter_04/case34-induction_lamp/induction_lamp_demo.c
#include <stdio.h>
#include <unistd.h>
#include "ohos_init.h"
#include "cmsis_os2.h"
#include "wifiiot_gpio.h"
#include "wifiiot_gpio_ex.h"
```

（4）实现人体感应灯功能，代码如下：

```
//applications/sample/wifi-iot/app/ohos_50/chapter_04/case34-induction_lamp/induction_lamp_demo.c
```

```c
void init(void)
{
    GpioInit();
    //绿色 LED 的 IO 为输出状态
    IoSetFunc(WIFI_IOT_IO_NAME_GPIO_10, WIFI_IOT_IO_FUNC_GPIO_10_GPIO);
    GpioSetDir(WIFI_IOT_IO_NAME_GPIO_10, WIFI_IOT_GPIO_DIR_OUT);
}
static void induction_lamp_thread(void * arg)
{
    (void)arg;
    init();
    wifiiotGpioValue rel = 0;
    while (1)
    {
        GpioGetInputVal(WIFI_IOT_IO_NAME_GPIO_7, &rel);
        GpioSetOutputVal(WIFI_IOT_IO_NAME_GPIO_10, (int)rel);
        usleep(20 * 1000);
    }
}
static void entry(void)
{
    osThreadAttr_t attr;
    attr.name = "induction_lamp_thread";
    attr.attr_bits = 0U;
    attr.cb_mem = NULL;
    attr.cb_size = 0U;
    attr.stack_mem = NULL;
    attr.stack_size = 4096;
    attr.priority = osPriorityNormal;
    if (osThreadNew(induction_lamp_thread, NULL, &attr) == NULL)
    {
        printf("[induction_lamp_thread] Failed to create induction_lamp_thread!\n");
    }
}
APP_FEATURE_INIT(entry);
```

(5) 创建并编写模块,构建脚本 BUILD.gn,代码如下:

```
//applications/sample/wifi-iot/app/ohos_50/chapter_04/case34-induction_lamp/BUILD.gn
static_library("ch_04_induction_lamp") {
    sources = [
        "induction_lamp_demo.c",
    ]
    include_dirs = [
        "//kernel/liteos_m/components/cmsis/2.0",
```

```
        "//base/iot_hardware/interfaces/kits/wifiiot_lite",
    ]
}
```

（6）将应用模块配置到应用子系统，代码如下：

```
//applications/sample/wifi-iot/app/BUILD.gn
import("//build/lite/config/component/lite_component.gni")
lite_component("app") {
    features = [
        "ohos_50/chapter_04/case34-induction_lamp:ch_04_induction_lamp",
    ]
}
```

（7）测试：编译应用模块，将固件烧写到开发板，复位开发板，将手在人体感应器上方晃动，观察 LED 是否发光，如图 4-4 所示。

图 4-4　人体感应灯的运行效果

4.4　案例 35：智能红外报警器

本案例为智能红外报警器，实现红外监测、报警、警示功能。核心模块为人体红外感应器、LED 和蜂鸣器。

将主控板插到母板上，再将人体红外感应板插到母板上，将蜂鸣器的 GND、I/O、VCC 引脚分别与母板的 GND、P09、V33 相连，如图 4-5 所示。

开发步骤如下：

（1）在本章目录 chapter_04 中创建应用模块工程目录 case35-infrared_alarm。

图 4-5　红外报警器组装后的开发板

(2) 在应用模块工程目录 case35-infrared_alarm 中创建应用模块源码文件 infrared_alarm_demo.c。

(3) 引用必要的头文件，代码如下：

```c
//applications/sample/wifi-iot/app/ohos_50/chapter_04/case35-infrared_alarm/infrared_alarm_demo.c
#include <stdio.h>
#include <unistd.h>
#include "ohos_init.h"
#include "cmsis_os2.h"
#include "wifiiot_gpio.h"
#include "wifiiot_gpio_ex.h"
```

(4) 实现红外报警功能，代码如下：

```c
//applications/sample/wifi-iot/app/ohos_50/chapter_04/case35-infrared_alarm/infrared_alarm_demo.c
void init(void)
{
    GpioInit();
    IoSetFunc(WIFI_IOT_IO_NAME_GPIO_2, WIFI_IOT_IO_FUNC_GPIO_2_GPIO);
    GpioSetDir(WIFI_IOT_IO_NAME_GPIO_2, WIFI_IOT_GPIO_DIR_OUT);
    //绿色 LED 的 IO 为输出状态
    IoSetFunc(WIFI_IOT_IO_NAME_GPIO_10, WIFI_IOT_IO_FUNC_GPIO_10_GPIO);
    GpioSetDir(WIFI_IOT_IO_NAME_GPIO_10, WIFI_IOT_GPIO_DIR_OUT);
}

static void infrared_alarm_thread(void *arg)
{
    (void)arg;
```

```
    init();
    wifiiotGpioValue rel = 0;
    while (1)
    {
        GpioGetInputVal(WIFI_IOT_IO_NAME_GPIO_7, &rel);
        GpioSetOutputVal(WIFI_IOT_IO_NAME_GPIO_10, (int)rel);
        int beep = rel?0:1;
        GpioSetOutputVal(WIFI_IOT_IO_NAME_GPIO_2, beep);
        usleep(20 * 1000);
    }
}
static void entry(void)
{
    osThreadAttr_t attr;
    attr.name = "infrared_alarm_thread";
    attr.attr_bits = 0U;
    attr.cb_mem = NULL;
    attr.cb_size = 0U;
    attr.stack_mem = NULL;
    attr.stack_size = 4096;
    attr.priority = osPriorityNormal;

    if (osThreadNew(infrared_alarm_thread, NULL, &attr) == NULL)
    {
        printf("[infrared_alarm_thread] Failed to create infrared_alarm_thread!\n");
    }
}
APP_FEATURE_INIT(entry);
```

(5) 创建并编写模块,构建脚本 BUILD.gn,代码如下:

```
//applications/sample/wifi-iot/app/ohos_50/chapter_04/case35-infrared_alarm/BUILD.gn
static_library("ch_04_infrared_alarm") {
    sources = [
        "infrared_alarm_demo.c",
    ]
    include_dirs = [
        "//kernel/liteos_m/components/cmsis/2.0",
        "//base/iot_hardware/interfaces/kits/wifiiot_lite",
    ]
}
```

(6) 将应用模块配置到应用子系统,代码如下:

```
//applications/sample/wifi-iot/app/BUILD.gn
import("//build/lite/config/component/lite_component.gni")
```

```
lite_component("app") {
    features = [
        "ohos_50/chapter_04/case35 - infrared_alarm:ch_04_infrared_alarm",
    ]
}
```

（7）测试：编译应用模块，将固件烧写到开发板，复位开发板，将手在人体感应器上方晃动。观察 LED 是否发光及确认蜂鸣器是否发出声音，如图 4-6 所示。

图 4-6　红外报警器的运行效果

4.5　案例 36：智能火焰报警器

本案例为智能火焰报警器，实现火焰智能监测、报警、警示功能。核心模块为火焰探测器、LED 和蜂鸣器。

将主控板插到母板上，再将火焰探测器的 DO、GND、VCC 分别与母板的 P07、GND、V33 相连，将蜂鸣器的 GND、I/O、VCC 引脚分别与母板的 GND、P09、V33 相连，如图 4-7 所示。

图 4-7　火焰报警器组装后的开发板

开发步骤如下：

(1) 在本章目录 chapter_04 中创建应用模块工程目录 case36-flame_alarm。

(2) 在应用模块工程目录 case36-flame_alarm 中创建应用模块源码文件 flame_alarm_demo.c。

(3) 引用必要的头文件，代码如下：

```c
//applications/sample/wifi-iot/app/ohos_50/chapter_04/case36-flame_alarm/flame_alarm_demo.c
#include <stdio.h>
#include <unistd.h>
#include "ohos_init.h"
#include "cmsis_os2.h"
#include "wifiiot_gpio.h"
#include "wifiiot_gpio_ex.h"
```

(4) 实现火焰报警器功能，代码如下：

```c
//applications/sample/wifi-iot/app/ohos_50/chapter_04/case36-flame_alarm/flame_alarm_demo.c
void init(void)
{
    GpioInit();
    IoSetFunc(WIFI_IOT_IO_NAME_GPIO_2, WIFI_IOT_IO_FUNC_GPIO_2_GPIO);
    GpioSetDir(WIFI_IOT_IO_NAME_GPIO_2, WIFI_IOT_GPIO_DIR_OUT);
}

static void infrared_alarm_thread(void *arg)
{
    (void)arg;

    init();
    wifiiotGpioValue rel = 0;
    while (1)
    {
        //读取当前是否有火焰
        GpioGetInputVal(WIFI_IOT_IO_NAME_GPIO_7, &rel);
        //当有火焰时读取低电平(0),蜂鸣器发出警报;当无火焰时读取高电平(1),蜂鸣器不发出警报
        GpioSetOutputVal(WIFI_IOT_IO_NAME_GPIO_2, (int)rel);
        usleep(20 * 1000);
    }
```

```
}
static void entry(void)
{

    osThreadAttr_t attr;
    attr.name = "flame_alarm_thread";
    attr.attr_bits = 0U;
    attr.cb_mem = NULL;
    attr.cb_size = 0U;
    attr.stack_mem = NULL;
    attr.stack_size = 4096;
    attr.priority = osPriorityNormal;
    if (osThreadNew(infrared_alarm_thread, NULL, &attr) == NULL)
    {
        printf("[flame_alarm_thread] Failed to create flame_alarm_thread!\n");
    }
}
APP_FEATURE_INIT(entry);
```

(5) 创建并编写模块,构建脚本 BUILD.gn,代码如下:

```
//applications/sample/wifi-iot/app/ohos_50/chapter_04/case36-flame_alarm/BUILD.gn
static_library("ch_04_flame_alarm") {
    sources = [
        "flame_alarm_demo.c",
    ]
    include_dirs = [
        "//kernel/liteos_m/components/cmsis/2.0",
        "//base/iot_hardware/interfaces/kits/wifiiot_lite",
    ]
}
```

(6) 将应用模块配置到应用子系统,代码如下:

```
//applications/sample/wifi-iot/app/BUILD.gn
import("//build/lite/config/component/lite_component.gni")
lite_component("app") {
    features = [
        "ohos_50/chapter_04/case36-flame_alarm:ch_04_flame_alarm",
    ]
}
```

（7）测试：编译应用模块，将固件烧写到开发板，复位开发板，将火焰对准火焰探测头。聆听蜂鸣器是否发出声音，如图 4-8 所示。

图 4-8　火焰报警器的运行效果

第 5 章 OpenHarmony 轻量系统物联网开发

本章将通过 13 个案例详细讲解 OpenHarmony 轻量系统在物联网应用中的开发技术，涉及的技术有 WiFi 组网、cJSON、网络通信和物联网。

5.1 WiFi 技术

无线保真（Wireless Fidelity，WiFi）在无线局域网的范畴是指"无线兼容性认证"，实际上是一种商业认证，同时也是一种无线联网技术，与蓝牙技术一样，同属于在办公室和家庭中使用的短距离无线技术。同蓝牙技术相比，它具备更高的传输速率，更远的传播距离，已经广泛应用于笔记本、手机、汽车等广大领域中。

WiFi 是无线局域网联盟的一个商标，该商标仅保障使用该商标的商品互相可以合作，与标准本身实际上没有关系，但因为 WiFi 主要采用 802.11b 协议，因此人们逐渐习惯用 WiFi 来称呼 802.11b 协议。从包含关系上来讲，WiFi 是 WLAN 的一个标准，WiFi 包含于 WLAN 中，属于采用 WLAN 协议中的一项新技术。

WiFi 是由无线接入点（Access Point，AP）、站点（Station，STA）等组成的无线网络。

（1）STA：每个连接到无线网络中的终端。

（2）AP：无线网络的创建者，是网络的中心节点。

STA 设备如图 5-1 所示。

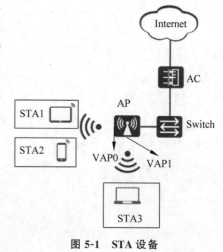

图 5-1 STA 设备

5.1.1 案例 37：STA 端点接入

本案例为 STA 端点接入，实现 STA 端点接入无线网络的功能。STA 端点接入的开发流程如下：

(1) 将待连接的 AP 属性配置到结构体 WifiDeviceConfig 类型变量中。
(2) 调用函数 EnableWifi()，启用 STA 模式。
(3) 调用函数 AddDeviceConfig()，配置热点信息，生成网络 ID。
(4) 调用函数 ConnectTo()，连接到指定的 AP 网络。
(5) 调用函数 netifapi_netif_find()，获取网络接口。
(6) 调用函数 netifapi_dhcp_start()，启动 DHCP 客户端，获取 IP 地址。

开发步骤如下：

(1) 创建本章源码存放目录 chapter_05。
(2) 在本章目录 chapter_05 中创建应用模块工程目录 case37-wifi_sta。
(3) 在应用模块工程目录 case37-wifi_sta 中创建应用模块源码文件 wifi_sta_demo.c。
(4) 引用必要的头文件，代码如下：

```c
//applications/sample/wifi-iot/app/ohos_50/chapter_05/case37-wifi_sta/wifi_sta_demo.c
#include <stdio.h>
#include <string.h>
#include <unistd.h>
#include "ohos_init.h"
#include "cmsis_os2.h"
#include "wifi_device.h"
#include "lwip/netifapi.h"
#include "lwip/api_shell.h"
```

(5) 实现 WiFi 连接功能，代码如下：

```c
//applications/sample/wifi-iot/app/ohos_50/chapter_05/case37-wifi_sta/wifi_sta_demo.c
static void wifi_sta_thread(void *arg)
{
    (void)arg;
    WifiErrorCode errCode;
    WifiDeviceConfig apConfig = {};
    int netId = -1;
    //配置将要连接的 AP 属性
    strcpy(apConfig.ssid, "Tenda_1F7390");              //网络名称
    strcpy(apConfig.preSharedKey, "nasoftmt");          //密码
    apConfig.securityType = WIFI_SEC_TYPE_PSK;          //加密类型

    errCode = EnableWifi();                             //启用 STA 模式
    errCode = AddDeviceConfig(&apConfig, &netId);       //配置热点信息，生成网络 ID
    printf("AddDeviceConfig: %d\r\n", errCode);
    errCode = ConnectTo(netId);                         //连接到指定的网络
    printf("ConnectTo(%d): %d\r\n", netId, errCode);
    usleep(3000 * 1000);                                //等待连接
    //获取网络接口，用于 IP 操作
```

```
      struct netif * iface = netifapi_netif_find("wlan0");
      if (iface)
      {
         //启动 DHCP 客户端,获取 IP 地址
         err_t ret = netifapi_dhcp_start(iface);
         printf("netifapi_dhcp_start: %d\r\n", ret);
         usleep(2000 * 1000);
         //netifapi_netif_common 用于以线程安全的方式调用所有与 netif 相关的 API
         //dhcp_clients_info_show 为 Shell API,用于展示 dhcp 客户端信息
         ret = netifapi_netif_common(iface, dhcp_clients_info_show, NULL);
         printf("netifapi_netif_common: %d\r\n", ret);
      }
      usleep(5000 * 1000);
      Disconnect();                                        //断开 WiFi 连接
      RemoveDevice(netId);                                 //移除 WiFi 热点的配置
      errCode = DisableWifi();                             //禁用 STA 模式
      printf("DisableWifi: %d\r\n", errCode);
   }
static void entry(void)
{
   osThreadAttr_t attr;
   attr.name = "wifi_sta_thread";
   attr.attr_bits = 0U;
   attr.cb_mem = NULL;
   attr.cb_size = 0U;
   attr.stack_mem = NULL;
   attr.stack_size = 4096;
   attr.priority = osPriorityNormal;
   if (osThreadNew(wifi_sta_thread, NULL, &attr) == NULL)
   {
      printf("[wifi_sta_thread] Failed to create wifi_sta_thread!\n");
   }
}
APP_FEATURE_INIT(entry);
```

(6) 创建并编写模块,构建脚本 BUILD.gn,代码如下:

```
//applications/sample/wifi-iot/app/ohos_50/chapter_05/case37-wifi_sta/BUILD.gn
static_library("ch_05_wifi_sta") {
   sources = [
      "wifi_sta_demo.c",
   ]
   include_dirs = [
      "//kernel/liteos_m/components/cmsis/2.0",
      "//base/iot_hardware/interfaces/kits/wifiiot_lite",
```

```
        "//vendor/hisi/hi3861/hi3861/third_party/lwip_sack/include/",
        "//foundation/communication/interfaces/kits/wifi_lite/wifiservice",
    ]
}
```

(7) 将应用模块配置到应用子系统,代码如下:

```
//applications/sample/wifi-iot/app/BUILD.gn
import("//build/lite/config/component/lite_component.gni")
lite_component("app") {
    features = [
        "ohos_50/chapter_05/case37-wifi_sta:ch_05_wifi_sta",
    ]
}
```

(8) 测试:编译应用模块,将固件烧写到开发板,复位开发板,将开发板与 IPOP 终端工具连接。观察 IPOP 终端工具窗口的打印信息,如图 5-2 所示。

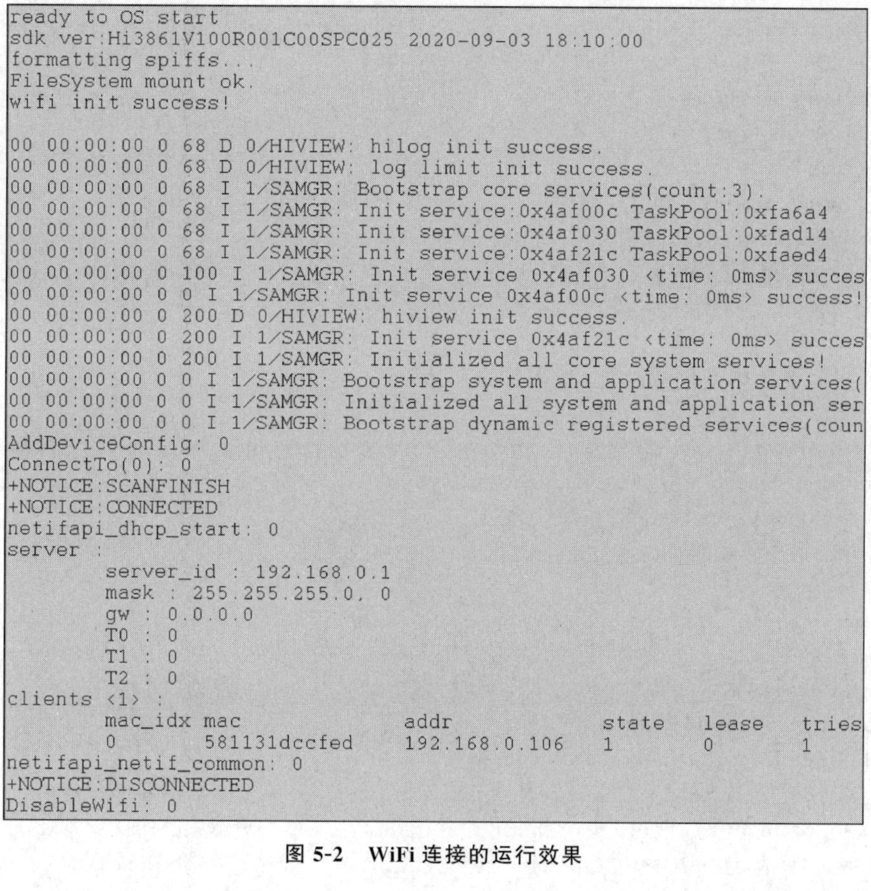

```
ready to OS start
sdk ver:Hi3861V100R001C00SPC025 2020-09-03 18:10:00
formatting spiffs...
FileSystem mount ok.
wifi init success!

00 00:00:00 0 68 D 0/HIVIEW: hilog init success.
00 00:00:00 0 68 D 0/HIVIEW: log limit init success.
00 00:00:00 0 68 I 1/SAMGR: Bootstrap core services(count:3).
00 00:00:00 0 68 I 1/SAMGR: Init service:0x4af00c TaskPool:0xfa6a4
00 00:00:00 0 68 I 1/SAMGR: Init service:0x4af030 TaskPool:0xfad14
00 00:00:00 0 68 I 1/SAMGR: Init service:0x4af21c TaskPool:0xfaed4
00 00:00:00 0 100 I 1/SAMGR: Init service 0x4af030 <time: 0ms> succes
00 00:00:00 0 0 I 1/SAMGR: Init service 0x4af00c <time: 0ms> success!
00 00:00:00 0 200 D 0/HIVIEW: hiview init success.
00 00:00:00 0 200 I 1/SAMGR: Init service 0x4af21c <time: 0ms> succes
00 00:00:00 0 200 I 1/SAMGR: Initialized all core system services!
00 00:00:00 0 0 I 1/SAMGR: Bootstrap system and application services(
00 00:00:00 0 0 I 1/SAMGR: Initialized all system and application ser
00 00:00:00 0 0 I 1/SAMGR: Bootstrap dynamic registered services(coun
AddDeviceConfig: 0
ConnectTo(0): 0
+NOTICE:SCANFINISH
+NOTICE:CONNECTED
netifapi_dhcp_start: 0
server
        server_id : 192.168.0.1
        mask : 255.255.255.0, 0
        gw : 0.0.0.0
        T0 : 0
        T1 : 0
        T2 : 0
clients <1> :
        mac_idx mac             addr            state   lease   tries
        0       581131dccfed    192.168.0.106   1       0       1
netifapi_netif_common: 0
+NOTICE:DISCONNECTED
DisableWifi: 0
```

图 5-2　WiFi 连接的运行效果

5.1.2 案例38：AP站点创建

本案例为AP站点创建，实现无线网络站点AP创建的功能。AP站点的创建流程如下：

(1) 将AP属性(账号、密码和加密类型)配置到结构体hi_wifi_softap_config类型变量中。

(2) 调用函数hi_wifi_softap_start()，启动AP功能。

(3) 调用函数netifapi_netif_find()，获取网络接口。

(4) 调用函数netifapi_netif_set_addr()，配置网络参数。

开发步骤如下：

(1) 在本章目录chapter_05中创建应用模块工程目录case38-wifi_ap。

(2) 在应用模块工程目录case38-wifi_ap中创建应用模块源码文件wifi_ap_demo.c。

(3) 引用必要的头文件，代码如下：

```
//applications/sample/wifi-iot/app/ohos_50/chapter_05/case38-wifi_ap/wifi_ap_demo.c
#include <stdio.h>
#include <string.h>
#include <unistd.h>
#include "ohos_init.h"
#include "cmsis_os2.h"
#include "hi_wifi_api.h"
#include "lwip/ip_addr.h"
#include "lwip/netifapi.h"
```

(4) 实现AP热点创建功能，代码如下：

```
//applications/sample/wifi-iot/app/ohos_50/chapter_05/case38-wifi_ap/wifi_ap_demo.c
static struct netif *g_lwip_netif = NULL;
static void wifi_ap_thread(void *arg)
{
    (void)arg;
    int ret;
    errno_t rc;
    char ifname[WIFI_IFNAME_MAX_SIZE + 1] = {0};      //保存返回的接口名字
    int len = sizeof(ifname);
    hi_wifi_softap_config hapd_conf = {0};
    ip4_addr_t st_gw;                                  //更新后网关的IP地址
    ip4_addr_t st_ipaddr;                              //更新后设备的IP地址
    ip4_addr_t st_netmask;                             //更新后的子网掩码
```

```c
    rc = memcpy_s(hapd_conf.ssid, HI_WIFI_MAX_SSID_LEN + 1, "wifi_sta_test", 16);
    if (rc != EOK)
    {
        return;
    }
    //设置密码(密码不少于8位)
    rc = memcpy_s(hapd_conf.key, HI_WIFI_MAX_KEY_LEN + 1, "12345678", 8);
    if (rc != EOK)
    {
        return;
    }
    //设置加密模式
    hapd_conf.authmode = HI_WIFI_SECURITY_WPA2PSK;
    hapd_conf.channel_num = 1;
    //启动AP功能
    ret = hi_wifi_softap_start(&hapd_conf, ifname, &len);
    if (ret != HISI_OK)
    {
        printf("hi_wifi_softap_start\n");
        return;
    }

    g_lwip_netif = netifapi_netif_find(ifname);
    if (g_lwip_netif == NULL)
    {
        printf("%s: get netif failed\n", __FUNCTION__);
        return;
    }
    IP4_ADDR(&st_gw, 192, 168, 56, 1);
    IP4_ADDR(&st_ipaddr, 192, 168, 56, 1);
    IP4_ADDR(&st_netmask, 255, 255, 255, 0);
    netifapi_netif_set_addr(g_lwip_netif, &st_ipaddr, &st_netmask, &st_gw);
    netifapi_dhcps_start(g_lwip_netif, 0, 0);
}
static void entry(void)
{
    osThreadAttr_t attr;
    attr.name = "wifi_ap_thread";
    attr.attr_bits = 0U;
    attr.cb_mem = NULL;
    attr.cb_size = 0U;
    attr.stack_mem = NULL;
```

```
    attr.stack_size = 4096;
    attr.priority = osPriorityNormal;

    if (osThreadNew(wifi_ap_thread, NULL, &attr) == NULL)
    {
        printf("[wifi_ap_thread] Failed to create wifi_ap_thread!\n");
    }
}

APP_FEATURE_INIT(entry);
```

(5) 创建并编写模块,构建脚本 BUILD.gn,代码如下:

```
//applications/sample/wifi-iot/app/ohos_50/chapter_05/case38-wifi_ap/BUILD.gn
static_library("ch_05_wifi_ap") {
    sources = [
        "wifi_ap_demo.c",
    ]
    include_dirs = [
        "//kernel/liteos_m/components/cmsis/2.0",
        "//base/iot_hardware/interfaces/kits/wifiiot_lite",
        "//vendor/hisi/hi3861/hi3861/third_party/lwip_sack/include/",
        "//foundation/communication/interfaces/kits/wifi_lite/wifiservice",
    ]
}
```

(6) 将应用模块配置到应用子系统,代码如下:

```
//applications/sample/wifi-iot/app/BUILD.gn
import("//build/lite/config/component/lite_component.gni")
lite_component("app") {
    features = [
        "ohos_50/chapter_05/case38-wifi_ap:ch_05_wifi_ap",
    ]
}
```

(7) 测试:编译应用模块,将固件烧写到开发板,复位开发板,将开发板与 IPOP 终端工具连接,用计算机连接到开发板的 AP 热点。观察 IPOP 终端工具窗口的打印信息,如图 5-3 所示。

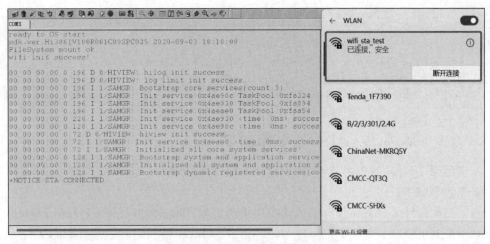

图 5-3　AP 热点创建并运行的效果

5.2　cJSON

cJSON 是一个超轻巧、携带方便、单文件、简单的可以作为 ANSI-C 标准的 JSON 解析器。cJSON 在 third_party\cJSON 下,其中主要包括两个文件 cjson.c 及 cjson.h,其中在 cjson.h 文件中包含了对于 JSON 格式的结构体定义及一些操作 JSON 格式的功能函数,包括创建 JSON、向 JSON 格式中添加数据(如数字、字符、布尔值等)、读取 JSON 格式、将 JSON 格式转换为字符串等,而在 cjson.c 文件中包含了功能函数的具体实现。下面了解一下 cJSON 的结构体:

```
//third_party/cJSON/cJSON.h
/* cJSON 结构体 */
typedef struct cJSON
{
    /* next 用于指向后一个对象,prev 用于指向前一个对象,从而形成双向对象链.next/prev 允许
遍历数组/对象链.或者使用 GetArraySize/GetArrayItem/ GetObjectItem */
    struct cJSON * next;
    struct cJSON * prev;
    /* 数组或对象项将具有指向数组/对象中项链的子指针 */
    struct cJSON * child;
    /* 项目的类型 */
    int type;
    /* 如果类型等于 cJSON_string,类型等于 c JSON_Raw,则为项的字符串 */
    char * valuestring;
    /* 写入 valueint 时已弃用,应改用 cJSON_SetNumberValue */
    int valueint;
```

```c
/* 项目编号,如果类型等于cJSON_number */
double valuedouble;

/* 如果此项是对象的子项或在对象的子项列表中,则为该项的名称字符串 */
char *string;
} cJSON;
```

5.2.1 案例39：cJSON 对象封装

本案例为 cJSON 对象封装,实现 cJSON 对象封装功能。

开发步骤如下：

(1) 在本章目录 chapter_05 中创建应用模块工程目录 case39-cjson_create_object。

(2) 在应用模块工程目录 case39-cjson_create_object 中创建应用模块源码文件 cjson_create_object_demo.c。

(3) 引用必要的头文件,代码如下：

```c
//applications/sample/wifi-iot/app/ohos_50/chapter_05/case39-cjson_create_object/cjson_create_object_demo.c
#include <stdio.h>
#include <string.h>
#include <unistd.h>
#include "cJSON.h"
#include "ohos_init.h"
#include "cmsis_os2.h"
```

(4) 实现 cJSON 对象封装功能,代码如下：

```c
//applications/sample/wifi-iot/app/ohos_50/chapter_05/case39-cjson_create_object/cjson_create_object_demo.c
static void cjson_create_object_thread(void *arg)
{
    (void)arg;
    sleep(1);
    cJSON *weather;
    weather = cJSON_CreateObject();
    //在cJSON对象中添加键-值对(值为整型)
    cJSON_AddNumberToObject(weather,"place",110108); //110108为北京市海淀区的地址代码
    //在cJSON对象中添加键-值对(值为字符串)
    cJSON_AddStringToObject(weather,"wc","sunny");
    //在cJSON对象中添加键-值对(值为浮点型)
    cJSON_AddNumberToObject(weather,"tempreture",27.5);
    //将JSON形式打印成正常字符串形式
    char *out = cJSON_Print(weather);
```

```
        printf("%s\n", out);
        //释放内存
        cJSON_Delete(weather);
}
static void entry(void)
{
    osThreadAttr_t attr;
    attr.name = "cjson_create_object_thread";
    attr.attr_bits = 0U;
    attr.cb_mem = NULL;
    attr.cb_size = 0U;
    attr.stack_mem = NULL;
    attr.stack_size = 10240;
    attr.priority = osPriorityNormal;

    if (osThreadNew(cjson_create_object_thread, NULL, &attr) == NULL)
    {
        printf("[cjson_create_object_thread] Failed to create cjson_create_object_thread!\n");
    }
}
APP_FEATURE_INIT(entry);
```

(5) 创建并编写模块,构建脚本 BUILD.gn,代码如下:

```
//applications/sample/wifi-iot/app/ohos_50/chapter_05/case39-cjson_create_object/BUILD.gn
static_library("ch_05_cjson_create_object") {
    sources = [
        "cjson_create_object_demo.c",
    ]
    include_dirs = [
        "//third_party/cJSON",
        "//kernel/liteos_m/components/cmsis/2.0",
        "//base/iot_hardware/interfaces/kits/wifiiot_lite",
    ]
}
```

(6) 将应用模块配置到应用子系统,代码如下:

```
//applications/sample/wifi-iot/app/BUILD.gn
import("//build/lite/config/component/lite_component.gni")
lite_component("app") {
    features = [
```

```
            "ohos_50/chapter_05/case39-cjson_create_object:ch_05_cjson_create_object",
    ]
}
```

(7) 测试：编译应用模块，将固件烧写到开发板，复位开发板，将开发板与 IPOP 终端工具连接。观察 IPOP 终端工具窗口的打印信息，如图 5-4 所示。

```
ready to OS start
sdk ver:Hi3861V100R001C00SPC025 2020-09-03 18:10:00
formatting spiffs...
FileSystem mount ok.
wifi init success!

00 00:00:00 0 196 D 0/HIVIEW: hilog init success.
00 00:00:00 0 196 D 0/HIVIEW: log limit init success.
00 00:00:00 0 196 I 1/SAMGR: Bootstrap core services(count:3).
00 00:00:00 0 196 I 1/SAMGR: Init service:0x4b2fa0 TaskPool:0xfa224
00 00:00:00 0 196 I 1/SAMGR: Init service:0x4b2fc4 TaskPool:0xfa894
00 00:00:00 0 196 I 1/SAMGR: Init service:0x4b31c4 TaskPool:0xfaa54
00 00:00:00 0 228 I 1/SAMGR: Init service 0x4b2fc4 <time: 0ms> succes
00 00:00:00 0 128 I 1/SAMGR: Init service 0x4b2fa0 <time: 0ms> succes
00 00:00:00 0 72 D 0/HIVIEW: hiview init success.
00 00:00:00 0 72 I 1/SAMGR: Init service 0x4b31c4 <time: 0ms> success
00 00:00:00 0 72 I 1/SAMGR: Initialized all core system services!
00 00:00:00 0 128 I 1/SAMGR: Bootstrap system and application service
00 00:00:00 0 128 I 1/SAMGR: Initialized all system and application s
00 00:00:00 0 128 I 1/SAMGR: Bootstrap dynamic registered services(co
{
        "place":        110108,
        "wc":   "sunny",
        "tempreture":    27.5
}
```

图 5-4 cJSON 对象封装的运行效果

5.2.2 案例 40：cJSON 对象解析

本案例为 cJSON 对象解析，实现 JSON 格式字符串解析为 cJSON 对象功能。

开发步骤如下：

(1) 在本章目录 chapter_05 中创建应用模块工程目录 case40-cjson_parse_object。

(2) 在应用模块工程目录 case40-cjson_parse_object 中创建应用模块源码文件 cjson_parse_object_demo.c。

(3) 引用必要的头文件，代码如下：

```
//applications/sample/wifi-iot/app/ohos_50/chapter_05/case40-cjson_parse_object/cjson_parse_object_demo.c
#include <stdio.h>
#include <string.h>
#include <unistd.h>
#include "cJSON.h"
#include "ohos_init.h"
#include "cmsis_os2.h"
```

(4) 实现 cJSON 对象封装功能，代码如下：

```c
//applications/sample/wifi-iot/app/ohos_50/chapter_05/case40-cjson_parse_object/cjson_parse_object_demo.c
static void cjson_parse_object_thread(void *arg)
{
    (void)arg;
    sleep(1);
    cJSON *json, *json_place, *json_wc, *json_temp;
    char *out = "{\"place\":110108,\"wc\":\"sunny\",\"temp\":27.5}";
    json = cJSON_Parse(out);//解析成 JSON 形式
    json_place = cJSON_GetObjectItem(json, "place");//获取键值内容
    json_wc = cJSON_GetObjectItem(json, "wc");
    json_temp = cJSON_GetObjectItem(json, "temp");
    printf("place:%d,wc:%s,temp:%.1f\n", json_place->valueint, json_wc->valuestring, json_temp->valuedouble);

    cJSON_Delete(json); //释放内存
}
static void entry(void)
{
    osThreadAttr_t attr;
    attr.name = "cjson_parse_object_thread";
    attr.attr_bits = 0U;
    attr.cb_mem = NULL;
    attr.cb_size = 0U;
    attr.stack_mem = NULL;
    attr.stack_size = 10240;
    attr.priority = osPriorityNormal;
    if (osThreadNew(cjson_parse_object_thread, NULL, &attr) == NULL)
    {
        printf("[cjson_parse_object_thread] Failed to create cjson_parse_object_thread!\n");
    }
}
APP_FEATURE_INIT(entry);
```

(5) 创建并编写模块，构建脚本 BUILD.gn，代码如下：

```
//applications/sample/wifi-iot/app/ohos_50/chapter_05/case40-cjson_parse_object/BUILD.gn
static_library("ch_05_cjson_parse_object") {
    sources = [
        "cjson_parse_object_demo.c",
    ]
    include_dirs = [
        "//third_party/cJSON",
```

```
        "//kernel/liteos_m/components/cmsis/2.0",
        "//base/iot_hardware/interfaces/kits/wifiiot_lite",
    ]
}
```

(6)将应用模块配置到应用子系统,代码如下:

```
//applications/sample/wifi-iot/app/BUILD.gn
import("//build/lite/config/component/lite_component.gni")
lite_component("app") {
    features = [
        "ohos_50/chapter_05/case40-cjson_parse_object:ch_05_cjson_parse_object",
    ]
}
```

(7)测试:编译应用模块,将固件烧写到开发板,复位开发板,将开发板与IPOP终端工具连接。观察IPOP终端工具窗口的打印信息,如图5-5所示。

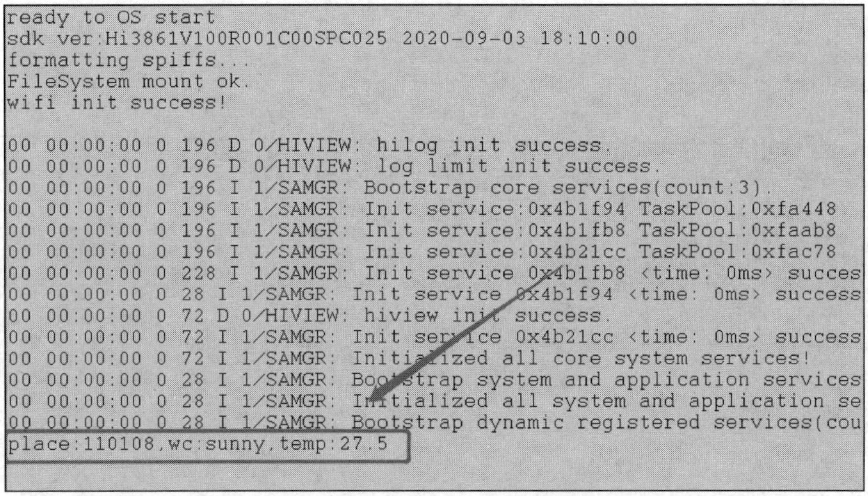

```
ready to OS start
sdk ver:Hi3861V100R001C00SPC025 2020-09-03 18:10:00
formatting spiffs...
FileSystem mount ok.
wifi init success!
00 00:00:00 0 196 D 0/HIVIEW: hilog init success.
00 00:00:00 0 196 D 0/HIVIEW: log limit init success.
00 00:00:00 0 196 I 1/SAMGR: Bootstrap core services(count:3).
00 00:00:00 0 196 I 1/SAMGR: Init service:0x4b1f94 TaskPool:0xfa448
00 00:00:00 0 196 I 1/SAMGR: Init service:0x4b1fb8 TaskPool:0xfaab8
00 00:00:00 0 196 I 1/SAMGR: Init service:0x4b21cc TaskPool:0xfac78
00 00:00:00 0 228 I 1/SAMGR: Init service 0x4b1fb8 <time: 0ms> succes
00 00:00:00 0 28 I 1/SAMGR: Init service 0x4b1f94 <time: 0ms> success
00 00:00:00 0 72 D 0/HIVIEW: hiview init success.
00 00:00:00 0 72 I 1/SAMGR: Init service 0x4b21cc <time: 0ms> success
00 00:00:00 0 72 I 1/SAMGR: Initialized all core system services!
00 00:00:00 0 28 I 1/SAMGR: Bootstrap system and application services
00 00:00:00 0 28 I 1/SAMGR: Initialized all system and application se
00 00:00:00 0 28 I 1/SAMGR: Bootstrap dynamic registered services(cou
place:110108,wc:sunny,temp:27.5
```

图 5-5 cJSON 对象解析的运行效果

5.2.3 案例41:cJSON 数组封装

本案例为 cJSON 数组封装,实现 cJSON 数组封装功能。
开发步骤如下:
(1)在本章目录 chapter_05 中创建应用模块工程目录 case41-cjson_create_array。
(2)在应用模块工程目录 case41-cjson_create_array 中创建应用模块源码文件 cjson_create_array_demo.c。
(3)引用必要的头文件,代码如下:

```c
//applications/sample/wifi-iot/app/ohos_50/chapter_05/case41-cjson_create_array/cjson_create_array_demo.c
#include <stdio.h>
#include <string.h>
#include <unistd.h>
#include "cJSON.h"
#include "ohos_init.h"
#include "cmsis_os2.h"
```

(4) 实现 cJSON 数组封装功能，代码如下：

```c
//applications/sample/wifi-iot/app/ohos_50/chapter_05/case41-cjson_create_array/cjson_create_array_demo.c
 static void cjson_create_array_thread(void *arg)
{
    (void)arg;
    sleep(1);
    cJSON *weather;
    weather = cJSON_CreateArray();
    cJSON_AddItemToArray(weather, cJSON_CreateNumber(110108));
    cJSON_AddItemToArray(weather, cJSON_CreateNumber(27.5));
    cJSON_AddItemToArray(weather, cJSON_CreateString("sunny"));
    //将JSON形式打印成正常字符串形式
    char *out = cJSON_Print(weather);
    printf("weather:%s\n", out);
    //释放内存
    cJSON_Delete(weather);
}
static void entry(void)
{
    osThreadAttr_t attr;
    attr.name = "cjson_create_array_thread";
    attr.attr_bits = 0U;
    attr.cb_mem = NULL;
    attr.cb_size = 0U;
    attr.stack_mem = NULL;
    attr.stack_size = 10240;
    attr.priority = osPriorityNormal;
    if (osThreadNew(cjson_create_array_thread, NULL, &attr) == NULL)
    {
        printf("[cjson_create_array_thread] Failed to create cjson_create_array_thread!\n");
    }
}
APP_FEATURE_INIT(entry);
```

(5) 创建并编写模块,构建脚本 BUILD.gn,代码如下:

```
//applications/sample/wifi-iot/app/ohos_50/chapter_05/case42-cjson_parse_array/BUILD.gn
static_library("ch_05_cjson_parse_array") {
    sources = [
        "cjson_parse_array_demo.c",
    ]
    include_dirs = [
        "//third_party/cJSON",
        "//kernel/liteos_m/components/cmsis/2.0",
        "//base/iot_hardware/interfaces/kits/wifiiot_lite",
    ]
}
```

(6) 将应用模块配置到应用子系统,代码如下:

```
//applications/sample/wifi-iot/app/BUILD.gn
import("//build/lite/config/component/lite_component.gni")
lite_component("app") {
    features = [
        "ohos_50/chapter_05/case42-cjson_parse_array:ch_05_cjson_parse_array",
    ]
}
```

(7) 测试:编译应用模块,将固件烧写到开发板,复位开发板,将开发板与 IPOP 终端工具连接。观察 IPOP 终端工具窗口的打印信息,如图 5-6 所示。

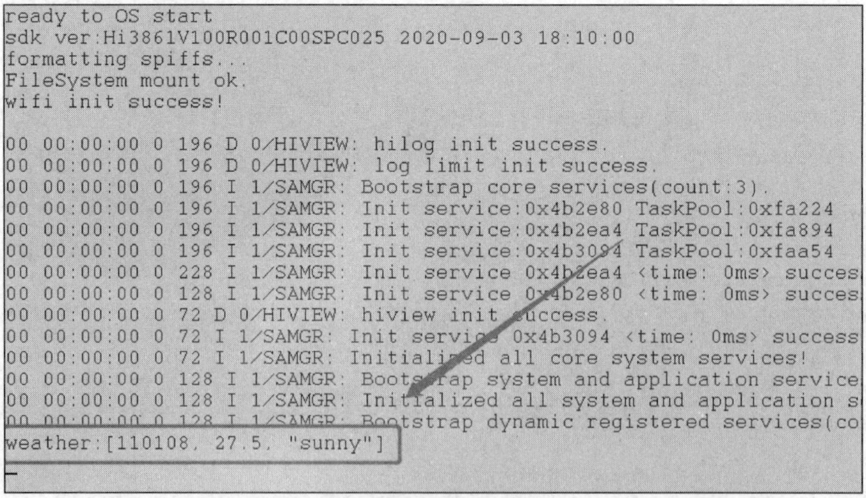

```
ready to OS start
sdk ver:Hi3861V100R001C00SPC025 2020-09-03 18:10:00
formatting spiffs...
FileSystem mount ok.
wifi init success!

00 00:00:00 0 196 D 0/HIVIEW: hilog init success.
00 00:00:00 0 196 D 0/HIVIEW: log limit init success.
00 00:00:00 0 196 I 1/SAMGR: Bootstrap core services(count:3).
00 00:00:00 0 196 I 1/SAMGR: Init service:0x4b2e80 TaskPool:0xfa224
00 00:00:00 0 196 I 1/SAMGR: Init service:0x4b2ea4 TaskPool:0xfa894
00 00:00:00 0 196 I 1/SAMGR: Init service:0x4b3094 TaskPool:0xfaa54
00 00:00:00 0 228 I 1/SAMGR: Init service 0x4b2ea4 <time: 0ms> succes
00 00:00:00 0 128 I 1/SAMGR: Init service 0x4b2e80 <time: 0ms> succes
00 00:00:00 0 72 D 0/HIVIEW: hiview init success.
00 00:00:00 0 72 I 1/SAMGR: Init service 0x4b3094 <time: 0ms> success
00 00:00:00 0 72 I 1/SAMGR: Initialized all core system services!
00 00:00:00 0 128 I 1/SAMGR: Bootstrap system and application service
00 00:00:00 0 128 I 1/SAMGR: Initialized all system and application s
00 00:00:00 0 128 I 1/SAMGR: Bootstrap dynamic registered services(co
weather:[110108, 27.5, "sunny"]
```

图 5-6　cJSON 数组封装的运行效果

5.2.4 案例 42：cJSON 数组解析

本案例为 cJSON 数组解析，实现 JSON 格式字符串解析为 cJSON 数组对象功能。
开发步骤如下：

(1) 在本章目录 chapter_05 中创建应用模块工程目录 case42-cjson_parse_array。

(2) 在应用模块工程目录 case42-cjson_parse_array 中创建应用模块源码文件 cjson_parse_array_demo.c。

(3) 引用必要的头文件，代码如下：

```c
//applications/sample/wifi-iot/app/ohos_50/chapter_05/case42-cjson_parse_array/cjson_parse_array_demo.c
#include <stdio.h>
#include <string.h>
#include <unistd.h>
#include "cJSON.h"
#include "ohos_init.h"
#include "cmsis_os2.h"
```

(4) 实现 cJSON 数组解析功能，代码如下：

```c
//applications/sample/wifi-iot/app/ohos_50/chapter_05/case42-cjson_parse_array/cjson_parse_array_demo.c
static void cjson_parse_array_thread(void *arg)
{
    (void)arg;
    sleep(1);
    cJSON *weather, *place, *temp, *wc;
    char *out = "[110108, 27.5, \"sunny\"]";
    weather = cJSON_Parse(out);
    place = cJSON_GetArrayItem(weather, 0);
    temp = cJSON_GetArrayItem(weather, 1);
    wc = cJSON_GetArrayItem(weather, 2);

    printf("place:%d,temp:%.1f,wc:%s\n", place->valueint, temp->valuedouble, wc->valuestring);
    //释放内存
    cJSON_Delete(weather);
}
static void entry(void)
{
    osThreadAttr_t attr;
```

```c
        attr.name = "cjson_parse_array_thread";
        attr.attr_bits = 0U;
        attr.cb_mem = NULL;
        attr.cb_size = 0U;
        attr.stack_mem = NULL;
        attr.stack_size = 10240;
        attr.priority = osPriorityNormal;
        if (osThreadNew(cjson_parse_array_thread, NULL, &attr) == NULL)
        {
            printf("[cjson_parse_array_thread] Failed to create cjson_parse_array_thread!\n");
        }
}
APP_FEATURE_INIT(entry);
```

(5) 创建并编写模块，构建脚本 BUILD.gn，代码如下：

```
//applications/sample/wifi-iot/app/ohos_50/chapter_05/case42-cjson_parse_array/BUILD.gn
static_library("ch_05_cjson_parse_array") {
    sources = [
        "cjson_parse_array_demo.c",
    ]
    include_dirs = [
        "//third_party/cJSON",
        "//kernel/liteos_m/components/cmsis/2.0",
        "//base/iot_hardware/interfaces/kits/wifiiot_lite",
    ]
}
```

(6) 将应用模块配置到应用子系统，代码如下：

```
//applications/sample/wifi-iot/app/BUILD.gn
import("//build/lite/config/component/lite_component.gni")
lite_component("app") {
    features = [
        "ohos_50/chapter_05/case42-cjson_parse_array:ch_05_cjson_parse_array",
    ]
}
```

(7) 测试：编译应用模块，将固件烧写到开发板，复位开发板，将开发板与 IPOP 终端工具连接。观察 IPOP 终端工具窗口的打印信息，如图 5-7 所示。

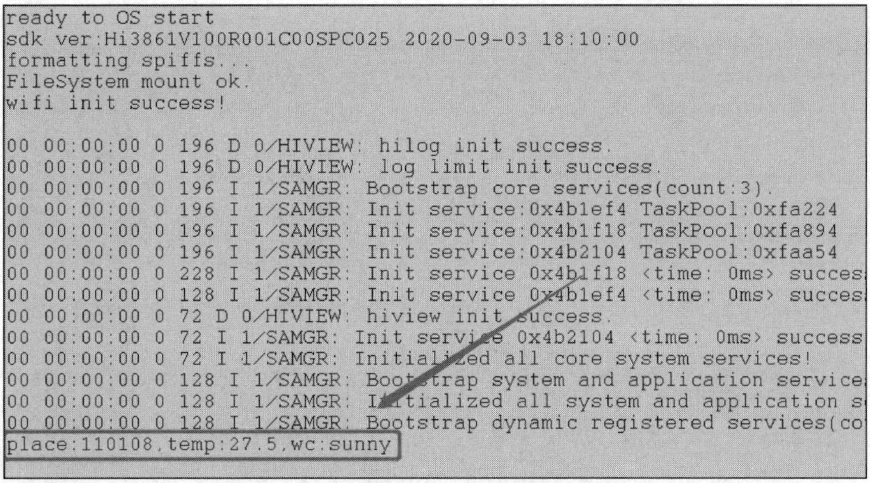

```
ready to OS start
sdk ver:Hi3861V100R001C00SPC025 2020-09-03 18:10:00
formatting spiffs...
FileSystem mount ok.
wifi init success!

00 00:00:00 0 196 D 0/HIVIEW: hilog init success.
00 00:00:00 0 196 D 0/HIVIEW: log limit init success.
00 00:00:00 0 196 I 1/SAMGR: Bootstrap core services(count:3).
00 00:00:00 0 196 I 1/SAMGR: Init service:0x4b1ef4 TaskPool:0xfa224
00 00:00:00 0 196 I 1/SAMGR: Init service:0x4b1f18 TaskPool:0xfa894
00 00:00:00 0 196 I 1/SAMGR: Init service:0x4b2104 TaskPool:0xfaa54
00 00:00:00 0 228 I 1/SAMGR: Init service 0x4b1f18 <time: 0ms> succes
00 00:00:00 0 128 I 1/SAMGR: Init service 0x4b1ef4 <time: 0ms> succes
00 00:00:00 0 72 D 0/HIVIEW: hiview init success.
00 00:00:00 0 72 I 1/SAMGR: Init service 0x4b2104 <time: 0ms> success
00 00:00:00 0 72 I 1/SAMGR: Initialized all core system services!
00 00:00:00 0 128 I 1/SAMGR: Bootstrap system and application service
00 00:00:00 0 128 I 1/SAMGR: Initialized all system and application s
00 00:00:00 0 128 I 1/SAMGR: Bootstrap dynamic registered services(co
place:110108,temp:27.5,wc:sunny
```

图 5-7 cJSON 数组解析的运行效果

5.3 网络通信

开源鸿蒙轻量系统提供了一套基于 LwIP 的网络通信 API。LwIP 是一个轻量级的 TCP/IP 协议栈。本节讲解基于 LwIP 的 TCP 与 UDP 客户端及服务器端的开发技术。

5.3.1 案例 43：UDP 客户端应用

本案例为 UDP 客户端应用，实现 UDP 客户端发送和接收数据功能。UDP 客户端的编程流程如下：

(1) 调用函数 socket()，创建套接字。

(2) 调用函数 sendto()，发送数据；调用函数 recvfrom()，接收数据。

(3) 调用函数 closesocket()，关闭套接字。

开发步骤如下：

(1) 在本章目录 chapter_05 中创建应用模块工程目录 case43-udp_client。

(2) 在应用模块工程目录 case43-udp_client 中创建应用模块源码文件 udp_client_demo.c。

(3) 引用必要的头文件并声明必要的全局变量，代码如下：

```
//applications/sample/wifi-iot/app/ohos_50/chapter_05/case43-udp_client/udp_client_demo.c
#include <stdio.h>
#include <unistd.h>
#include "ohos_init.h"
#include "cmsis_os2.h"
```

```c
#include "wifi_device.h"
#include "hi_wifi_api.h"
#include "lwip/ip_addr.h"
#include "lwip/netifapi.h"
#include "lwip/sockets.h"
#include "lwip/api_shell.h"
static char request[] = "Hello.I am Hi3861 udp_client.";
static char response[128] = "";
```

(4) 创建 send_to_udp_server 方法,实现 UDP 客户端的收发信息功能,代码如下:

```c
//applications/sample/wifi-iot/app/ohos_50/chapter_05/case43-udp_client/udp_client_demo.c
//参数 host 为 UDP 服务器端 IP 地址,参数 port 为 UDP 服务器端端口号
void send_to_udp_server(const char * host, unsigned short port)
{
    ssize_t retval = 0;
    int sockfd = socket(AF_INET, SOCK_DGRAM, 0);       //创建 Socket 套接字,并指定为数据报套接
                                                       //字(支持 UDP)
    struct sockaddr_in toAddr = {0};
    toAddr.sin_family = AF_INET;
    toAddr.sin_port = htons(port);                     //端口号,从主机字节序转换为网络字节序

    //将主机 IP 地址从"点分十进制"字符串转换为标准格式(32 位整数)
    if (inet_pton(AF_INET, host, &toAddr.sin_addr) <= 0)
    {
        printf("inet_pton failed!\r\n");
        goto do_cleanup;
    }
    printf("addr:%u\n",toAddr.sin_addr);
    //UDP Socket 是"无连接"的,因此每次发送都必须先指定目标主机和端口,主机可以是多播地址
    retval = sendto(sockfd, request, sizeof(request), 0, (struct sockaddr * )&toAddr, sizeof(toAddr));
    if (retval < 0)
    {
        printf("sendto failed!\r\n");
        goto do_cleanup;
    }
    printf("send UDP message {%s} %ld done!\r\n", request, retval);
    struct sockaddr_in fromAddr = {0};
    socklen_t fromLen = sizeof(fromAddr);
    //UDP Socket 是"无连接"的,因此每次接收时并不知道消息来自何处,通过 fromAddr 参数可以
    //得到发送方的信息(主机、端口号)
    retval = recvfrom(sockfd, &response, sizeof(response), 0, (struct sockaddr * )&fromAddr, &fromLen);
```

```c
    if (retval <= 0)
    {
        printf("recvfrom failed or abort, %ld, %d!\r\n", retval, errno);
        goto do_cleanup;
    }
    response[retval] = '\0';
    printf("recv UDP message {%s} %ld done!\r\n", response, retval);
    printf("peer info: ipaddr = %s, port = %d\r\n", inet_ntoa(fromAddr.sin_addr), ntohs(fromAddr.sin_port));
do_cleanup:
    printf("do_cleanup...\r\n");
    //关闭套接字
    closesocket(sockfd);
}
```

(5) 创建方法 connect_wifi,实现 WiFi 连接功能,代码如下:

```c
//applications/sample/wifi-iot/app/ohos_50/chapter_05/case43-udp_client/udp_client_demo.c
void connect_wifi(void)
{
    WifiErrorCode errCode;
    WifiDeviceConfig apConfig = {};
    int netId = -1;
    strcpy(apConfig.ssid, "Tenda_1F7390");
    strcpy(apConfig.preSharedKey, "nasoftmt");
    apConfig.securityType = WIFI_SEC_TYPE_PSK;

    errCode = EnableWifi();
    errCode = AddDeviceConfig(&apConfig, &netId);
    errCode = ConnectTo(netId);
    printf("ConnectTo(%d): %d\r\n", netId, errCode);
    usleep(1000 * 1000);
    struct netif * iface = netifapi_netif_find("wlan0");
    if (iface)
    {
        err_t ret = netifapi_dhcp_start(iface);
        printf("netifapi_dhcp_start: %d\r\n", ret);
        usleep(2000 * 1000);
        ret = netifapi_netif_common(iface, dhcp_clients_info_show, NULL);
        printf("netifapi_netif_common: %d\r\n", ret);
    }
}
```

(6) 创建 UDP 客户端任务,实现收发信息功能,代码如下:

```c
//applications/sample/wifi-iot/app/ohos_50/chapter_05/case43-udp_client/udp_client_demo.c
static void udp_client_thread(void *arg)
{
    (void)arg;
    sleep(1);

    connect_wifi();                    //连接 AP
    printf("udp client demo start\n");
    unsigned short port = 5002;
    send_to_udp_server("192.168.0.106", port);
}
static void entry(void)
{
    osThreadAttr_t attr;
    attr.name = "udp_client_thread";
    attr.attr_bits = 0U;
    attr.cb_mem = NULL;
    attr.cb_size = 0U;
    attr.stack_mem = NULL;
    attr.stack_size = 10240;
    attr.priority = osPriorityNormal;
    if (osThreadNew(udp_client_thread, NULL, &attr) == NULL)
    {
        printf("[udp_client_thread] Failed to create udp_client_thread!\n");
    }
}
APP_FEATURE_INIT(entry);
```

(7) 创建并编写模块,构建脚本 BUILD.gn,代码如下:

```
//applications/sample/wifi-iot/app/ohos_50/chapter_05/case43-udp_client/BUILD.gn
static_library("ch_05_udp_client") {
    sources = [
        "udp_client_demo.c",
    ]
    include_dirs = [
        "//kernel/liteos_m/components/cmsis/2.0",
        "//base/iot_hardware/interfaces/kits/wifiiot_lite",
        "//vendor/hisi/hi3861/hi3861/third_party/lwip_sack/include",
        "//foundation/communication/interfaces/kits/wifi_lite/wifiservice",
    ]
}
```

(8) 将应用模块配置到应用子系统,代码如下:

```
//applications/sample/wifi-iot/app/BUILD.gn
import("//build/lite/config/component/lite_component.gni")
lite_component("app") {
    features = [
        "ohos_50/chapter_05/case43-udp_client:ch_05_udp_client",
    ]
}
```

(9) 测试:编译应用模块,将固件烧写到开发板,复位开发板,将开发板与 IPOP 终端工具连接,并用 IPOP 服务工具创建 UDP 服务器,以便监听端口 5002。观察 IPOP 终端工具窗口的打印信息,如图 5-8 所示。

```
00 00:00:00 0 192 I 1/SAMGR: Bootstrap system and application service
00 00:00:00 0 192 I 1/SAMGR: Initialized all system and application s
00 00:00:00 0 192 I 1/SAMGR: Bootstrap dynamic registered services(co
ConnectTo(0): 0
+NOTICE:SCANFINISH
+NOTICE:CONNECTED
netifapi_dhcp_start: 0
server :
        server_id : 192.168.0.1
        mask : 255.255.255.0, 1
        gw : 192.168.0.1
        T0 : 86400
        T1 : 43200
        T2 : 75600
clients <1> :
        mac_idx mac             addr            state   lease   tries
        0       b01131f2147f    192.168.0.117   10      0       1
netifapi_netif_common: 0
udp client demo start
addr:1778428096
send UDP message {Hello.I am Hi3861 udp_client.} 30 done!
recv UDP message {Hello.I am Hi3861 udp_client.} 30 done!
peer info: ipaddr = 192.168.0.106, port = 5002
do_cleanup...
```

图 5-8 UDP 客户端应用(IPOP 终端工具)的运行效果

观察 IPOP UDP 服务器窗口的打印信息,如图 5-9 所示。

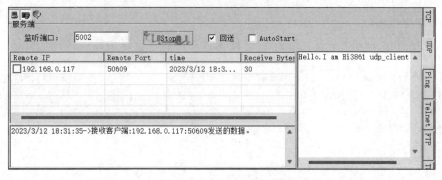

图 5-9 UDP 客户端应用(IPOP 服务器端)的运行效果

注意：如果在运行时连接不上 UDP 服务器端，则应在计算机设置中关闭防火墙，或开放所需端口。

5.3.2 案例 44：UDP 服务器端应用

本案例为 UDP 服务器端应用，实现 UDP 服务器端数据接收、发送功能。UDP 服务器端的编程流程如下：

(1) 调用函数 socket()，创建套接字。
(2) 调用函数 bind()，将服务器的 IP 地址、端口号与套接字进行绑定。
(3) 调用函数 recvfrom()，接收数据；调用函数 sendto()，发送数据。
(4) 调用函数 closesocket()，关闭套接字。

开发步骤如下：

(1) 在本章目录 chapter_05 中创建应用模块工程目录 case44-udp_server。
(2) 在应用模块工程目录 case44-udp_server 中创建应用模块源码文件 udp_server_demo.c。
(3) 引用必要的头文件，代码如下：

```c
//applications/sample/wifi-iot/app/ohos_50/chapter_05/case44-udp_server/udp_server_demo.c
#include <stdio.h>
#include <unistd.h>
#include "ohos_init.h"
#include "cmsis_os2.h"
#include "wifi_device.h"
#include "hi_wifi_api.h"
#include "lwip/ip_addr.h"
#include "lwip/netifapi.h"
#include "lwip/sockets.h"
#include "lwip/api_shell.h"
static char request[] = "Hello.I am Hi3861 udp_client.";
static char response[128] = "";
```

(4) 创建 udp_server 方法，实现 UDP 服务器端的收发信息功能，代码如下：

```c
//applications/sample/wifi-iot/app/ohos_50/chapter_05/case44-udp_server/udp_server_demo.c
//port 为服务器绑定端口
void udp_server(unsigned short port)
{
    ssize_t retval = 0;
    char message[128] = "";
```

```c
    int sockfd = socket(AF_INET, SOCK_DGRAM, 0);            //创建套接字
    struct sockaddr_in clientAddr = {0};
    socklen_t clientAddrLen = sizeof(clientAddr);
    struct sockaddr_in serverAddr = {0};
    serverAddr.sin_family = AF_INET;
    serverAddr.sin_port = htons(port);
    serverAddr.sin_addr.s_addr = htonl(INADDR_ANY);
    //将服务器的IP地址、端口号与套接字进行绑定
    retval = bind(sockfd, (struct sockaddr *)&serverAddr, sizeof(serverAddr));
    if (retval < 0)
    {
        printf("bind failed, %ld!\r\n", retval);
        goto do_cleanup;
    }
    printf("bind to port %d success!\r\n", port);
    while (1)
    {
        //接收来自客户端的数据
        retval = recvfrom(sockfd, message, sizeof(message), 0, (struct sockaddr *)&clientAddr, &clientAddrLen);
        if (retval < 0)
        {
            printf("recvfrom failed, %ld!\r\n", retval);
            goto do_cleanup;
        }
        printf("recv message {%s} %ld done!\r\n", message, retval);
        printf("peer info: ipaddr = %s, port = %d\r\n", inet_ntoa(clientAddr.sin_addr), ntohs(clientAddr.sin_port));
        //将数据发送到客户端
        retval = sendto(sockfd, message, strlen(message), 0, (struct sockaddr *)&clientAddr, sizeof(clientAddr));
        if (retval <= 0)
        {
            printf("send failed, %ld!\r\n", retval);
            goto do_cleanup;
        }
        printf("send message {%s} %ld done!\r\n", message, retval);
    }
do_cleanup:
    printf("do_cleanup...\r\n");
    //关闭套接字
    closesocket(sockfd);
}
```

(5) 创建方法 connect_wifi,实现 WiFi 连接功能,代码如下:

```c
//applications/sample/wifi-iot/app/ohos_50/chapter_05/case44-udp_server/udp_server_demo.c
void connect_wifi(void)
{
    WifiErrorCode errCode;
    WifiDeviceConfig apConfig = {};
    int netId = -1;
    strcpy(apConfig.ssid, "Tenda_1F7390");
    strcpy(apConfig.preSharedKey, "nasoftmt");
    apConfig.securityType = WIFI_SEC_TYPE_PSK;
    errCode = EnableWifi();
    errCode = AddDeviceConfig(&apConfig, &netId);
    errCode = ConnectTo(netId);
    printf("ConnectTo(%d): %d\r\n", netId, errCode);
    usleep(1000 * 1000);
    struct netif *iface = netifapi_netif_find("wlan0");
    if (iface)
    {
        err_t ret = netifapi_dhcp_start(iface);
        printf("netifapi_dhcp_start: %d\r\n", ret);
        usleep(2000 * 1000);
        ret = netifapi_netif_common(iface, dhcp_clients_info_show, NULL);
        printf("netifapi_netif_common: %d\r\n", ret);
    }
}
```

(6) 创建 UDP 服务器端任务,实现收发信息功能,代码如下:

```c
//applications/sample/wifi-iot/app/ohos_50/chapter_05/case44-udp_server/udp_server_demo.c
static void udp_server_thread(void *arg)
{
    (void)arg;
    sleep(1);
    connect_wifi();                        //连接 AP
    printf("udp client demo start\n");
    unsigned short port = 5001;
    udp_server(port);
}
static void entry(void)
{
    osThreadAttr_t attr;
    attr.name = "udp_server_thread";
    attr.attr_bits = 0U;
```

```
    attr.cb_mem = NULL;
    attr.cb_size = 0U;
    attr.stack_mem = NULL;
    attr.stack_size = 10240;
    attr.priority = osPriorityNormal;

    if (osThreadNew(udp_server_thread, NULL, &attr) == NULL)
    {
        printf("[udp_server_thread] Failed to create udp_server_thread!\n");
    }
}
APP_FEATURE_INIT(entry);
```

(7) 创建并编写模块，构建脚本 BUILD.gn，代码如下：

```
//applications/sample/wifi-iot/app/ohos_50/chapter_05/case44-udp_server/BUILD.gn
static_library("ch_05_udp_server") {
    sources = [
        "udp_server_demo.c",
    ]
    include_dirs = [
        "//kernel/liteos_m/components/cmsis/2.0",
        "//base/iot_hardware/interfaces/kits/wifiiot_lite",
        "//vendor/hisi/hi3861/hi3861/third_party/lwip_sack/include",
        "//foundation/communication/interfaces/kits/wifi_lite/wifiservice",
    ]
}
```

(8) 将应用模块配置到应用子系统，代码如下：

```
//applications/sample/wifi-iot/app/BUILD.gn
import("//build/lite/config/component/lite_component.gni")
lite_component("app") {
    features = [
        "ohos_50/chapter_05/case44-udp_server:ch_05_udp_server",
    ]
}
```

(9) 测试：编译应用模块，将固件烧写到开发板，复位开发板，将开发板与 IPOP 终端工具连接，然后用 IPOP 服务工具创建 UDP 客户端并向 UDP 服务器端发送信息。观察 IPOP UDP 客户端窗口的打印信息，如图 5-10 所示。

观察 IPOP 终端工具窗口的打印信息，如图 5-11 所示。

图 5-10 UDP 服务器端应用（IPOP 客户端）的运行效果

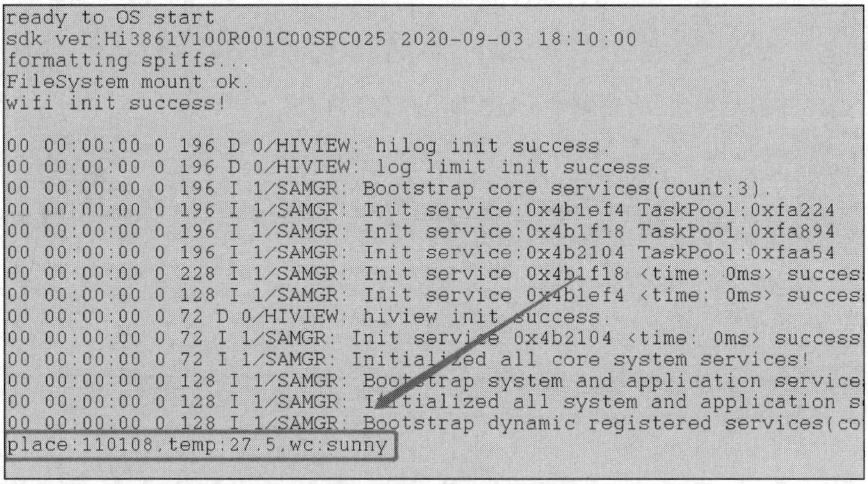

图 5-11 UDP 服务器端应用（IPOP 终端工具）的运行效果

5.3.3 案例 45：TCP 客户端应用

本案例为 TCP 客户端应用，实现 TCP 客户端与服务器端的连接建立与断开、数据发送与接收功能。TCP 客户端的编程流程如下：

（1）调用函数 socket()，创建套接字。

（2）调用函数 connect()，发送客户端连接请求。

（3）调用函数 send()，发送数据；调用函数 recv()，接收数据。

（4）调用函数 closesocket()，关闭套接字。

开发步骤如下：

（1）在本章目录 chapter_05 中创建应用模块工程目录 case45-tcp_client。

（2）在应用模块工程目录 case45-tcp_client 中创建应用模块源码文件 tcp_client_demo.c。

（3）引用必要的头文件并声明必要的全局变量，代码如下：

```c
//applications/sample/wifi-iot/app/ohos_50/chapter_05/case45-tcp_client/tcp_client_demo.c
#include <stdio.h>
#include <unistd.h>
#include "ohos_init.h"
#include "cmsis_os2.h"
#include "wifi_device.h"
#include "hi_wifi_api.h"
#include "lwip/ip_addr.h"
#include "lwip/netifapi.h"
#include "lwip/sockets.h"
#include "lwip/api_shell.h"
static char request[] = "Hello.I am from Hi3861 tcp_client.";
static char response[128] = "";
```

(4) 创建 conent_tcp_server 方法,实现 TCP 客户端的收发信息功能,代码如下:

```c
//applications/sample/wifi-iot/app/ohos_50/chapter_05/case45-tcp_client/tcp_client_demo.c
//参数 host 为 TCP 服务器端 IP 地址,参数 port 为 TCP 服务器端端口号
void conent_tcp_server(const char *host, unsigned short port)
{
    ssize_t retval = 0;
    int sockfd = socket(AF_INET, SOCK_STREAM, 0);
    struct sockaddr_in serverAddr = {0};
    serverAddr.sin_family = AF_INET;           //AF_INET 表示 IPv4 协议
    serverAddr.sin_port = htons(port);         //端口号,从主机字节序转换为网络字节序
    //将主机 IP 地址从"点分十进制"字符串转换为标准格式(32 位整数)
    if (inet_pton(AF_INET, host, &serverAddr.sin_addr) <= 0)
    {
        printf("inet_pton failed!\r\n");
        goto do_cleanup;
    }
    //尝试和目标主机建立连接,如果连接成功,则返回 0,如果连接失败,则返回 -1
    if (connect(sockfd, (struct sockaddr *)&serverAddr, sizeof(serverAddr)) < 0)
    {
        printf("connect failed!\r\n");
        goto do_cleanup;
    }
    printf("connect to server %s success!\r\n", host);
    //建立连接成功后,这个 TCP Socket 描述符——sockfd 就具有了"连接状态",可以对服务器发
    //送、接收数据
    retval = send(sockfd, request, sizeof(request), 0);
    if (retval < 0)
    {
```

```
      printf("send request failed!\r\n");
      goto do_cleanup;
   }
   printf("send request{%s} %ld to server done!\r\n", request, retval);
   retval = recv(sockfd, response, sizeof(response), 0);
   if (retval <= 0)
   {
      printf("send response from server failed or done, %ld!\r\n", retval);
      goto do_cleanup;
   }
   response[retval] = '\0';
   printf("recv response{%s} %ld from server done!\r\n", response, retval);
do_cleanup:
   printf("do_cleanup...\r\n");
   closesocket(sockfd);
}
```

(5) 创建方法 connect_wifi,实现 WiFi 连接功能,代码如下:

```
//applications/sample/wifi-iot/app/ohos_50/chapter_05/case45-tcp_client/tcp_client_demo.c
void connect_wifi(void)
{
   WifiErrorCode errCode;
   WifiDeviceConfig apConfig = {};
   int netId = -1;
   strcpy(apConfig.ssid, "Tenda_1F7390");
   strcpy(apConfig.preSharedKey, "nasoftmt");
   apConfig.securityType = WIFI_SEC_TYPE_PSK;
   errCode = EnableWifi();
   errCode = AddDeviceConfig(&apConfig, &netId);
   errCode = ConnectTo(netId);
   printf("ConnectTo(%d): %d\r\n", netId, errCode);
   usleep(1000 * 1000);
   struct netif *iface = netifapi_netif_find("wlan0");
   if (iface)
   {
      err_t ret = netifapi_dhcp_start(iface);
      printf("netifapi_dhcp_start: %d\r\n", ret);
      usleep(2000 * 1000);
      ret = netifapi_netif_common(iface, dhcp_clients_info_show, NULL);
      printf("netifapi_netif_common: %d\r\n", ret);
   }
}
```

(6) 创建 TCP 客户端任务，实现收发信息功能，代码如下：

```c
//applications/sample/wifi-iot/app/ohos_50/chapter_05/case45-tcp_client/tcp_client_demo.c
static void tcp_client_thread(void *arg)
{
    (void)arg;
    sleep(1);
    connect_wifi();
    sleep(1);
    printf("tcp client demo start\n");
    unsigned short port = 5001;
    conent_tcp_server("192.168.0.105", port);
}
static void entry(void)
{
    osThreadAttr_t attr;
    attr.name = "tcp_client_thread";
    attr.attr_bits = 0U;
    attr.cb_mem = NULL;
    attr.cb_size = 0U;
    attr.stack_mem = NULL;
    attr.stack_size = 10240;
    attr.priority = osPriorityNormal;
    if (osThreadNew(tcp_client_thread, NULL, &attr) == NULL)
    {
        printf("[tcp_client_thread] Failed to create tcp_client_thread!\n");
    }
}
APP_FEATURE_INIT(entry);
```

(7) 创建并编写模块，构建脚本 BUILD.gn，代码如下：

```
//applications/sample/wifi-iot/app/ohos_50/chapter_05/case45-tcp_client/BUILD.gn
static_library("ch_05_tcp_client") {
    sources = [
        "tcp_client_demo.c",
    ]
    include_dirs = [
        "//kernel/liteos_m/components/cmsis/2.0",
        "//base/iot_hardware/interfaces/kits/wifiiot_lite",
        "//vendor/hisi/hi3861/hi3861/third_party/lwip_sack/include",
        "//foundation/communication/interfaces/kits/wifi_lite/wifiservice",
    ]
}
```

(8) 将应用模块配置到应用子系统,代码如下:

```
//applications/sample/wifi-iot/app/BUILD.gn
import("//build/lite/config/component/lite_component.gni")
lite_component("app") {
   features = [
        "ohos_50/chapter_05/case45-tcp_client:ch_05_tcp_client",
   ]
}
```

(9) 测试:编译应用模块,将固件烧写到开发板,复位开发板,将开发板与 IPOP 终端工具连接,并用 IPOP 服务工具创建 TCP 服务器监听端口 5001。观察 IPOP 终端工具窗口的打印信息,如图 5-12 所示。

```
ConnectTo(0): 0
+NOTICE:SCANFINISH
+NOTICE:CONNECTED
netifapi_dhcp_start: 0
server :
        server_id : 192.168.0.1
        mask : 255.255.255.0, 0
        gw : 0.0.0.0
        T0 : 0
        T1 : 0
        T2 : 0
clients <1> :
        mac_idx mac             addr            state   lease   tries
        0       081131a0154a    192.168.0.121   1       0       1
netifapi_netif_common: 0
tcp client demo start
connect to server 192.168.0.105 success!
send request{Hello.I am from Hi3861 tcp_client.} 35 to server done!
recv response{Welcome to TcpSrv
} 19 from server done!
do_cleanup...
```

图 5-12　TCP 客户端应用(IPOP 终端工具)的运行效果

观察 IPOP TCP 服务器窗口的打印信息,如图 5-13 所示。

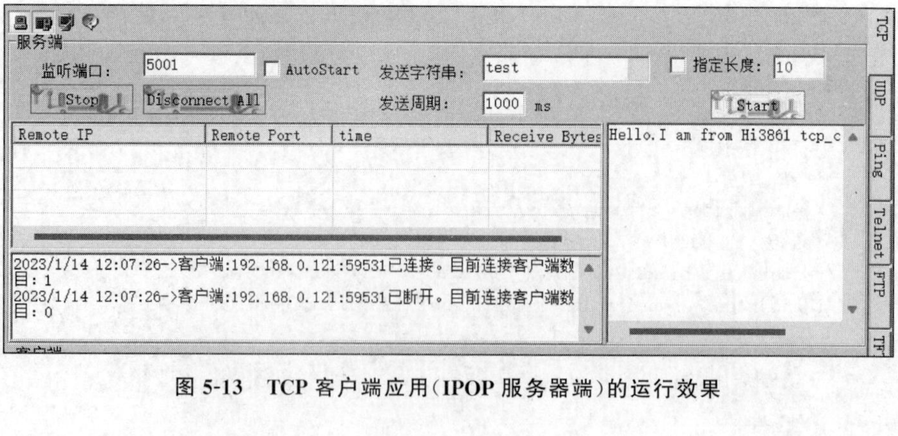

图 5-13　TCP 客户端应用(IPOP 服务器端)的运行效果

注意：如果在运行时连接不上 TCP 服务器端，则应在计算机设置中关闭防火墙，或开放所需端口。

5.3.4 案例46：TCP 服务器端应用

本案例为 TCP 服务器端应用，实现 TCP 服务器端收发数据功能。TCP 服务器端的编程流程如下：

(1) 调用函数 socket()，创建套接字。
(2) 调用函数 bind()，将服务器的 IP 地址、端口号与套接字进行绑定。
(3) 调用函数 listen()，将套接字设置为监听状态。
(4) 调用函数 accept()，阻塞等待客户端的连接请求。
(5) 调用函数 recv()，接收数据；调用函数 send()，发送数据。
(6) 调用函数 closesocket()，关闭套接字。

开发步骤如下：
(1) 在本章目录 chapter_05 中创建应用模块工程目录 case46-tcp_server。
(2) 在应用模块工程目录 case46-tcp_server 中创建应用模块源码文件 tcp_server_demo.c。
(3) 引用必要的头文件，代码如下：

```c
//applications/sample/wifi-iot/app/ohos_50/chapter_05/case46-tcp_server/tcp_server_demo.c
#include <stdio.h>
#include <unistd.h>
#include "ohos_init.h"
#include "cmsis_os2.h"
#include "wifi_device.h"
#include "hi_wifi_api.h"
#include "lwip/ip_addr.h"
#include "lwip/netifapi.h"
#include "lwip/sockets.h"
#include "lwip/api_shell.h"
```

(4) 创建 tcp_server 方法，实现 TCP 服务器端的收发信息功能，代码如下：

```c
//applications/sample/wifi-iot/app/ohos_50/chapter_05/case46-tcp_server/tcp_server_demo.c
void tcp_server(unsigned short port)
{
    char message[128] = "";
    ssize_t retval = 0;
```

```c
int backlog = 1;
int connfd = -1;
int sockfd = socket(AF_INET, SOCK_STREAM, 0);
struct sockaddr_in clientAddr = {0};
socklen_t clientAddrLen = sizeof(clientAddr);
struct sockaddr_in serverAddr = {0};
serverAddr.sin_family = AF_INET;
serverAddr.sin_port = htons(port);
//允许任意主机接入
serverAddr.sin_addr.s_addr = htonl(INADDR_ANY);
//绑定端口
retval = bind(sockfd, (struct sockaddr *)&serverAddr, sizeof(serverAddr));
if (retval < 0)
{
    printf("bind failed, %ld!\r\n", retval);
    goto do_cleanup;
}
printf("bind to port %d success!\r\n", port);
retval = listen(sockfd, backlog);              //开始监听
if (retval < 0)
{
    printf("listen failed!\r\n");
    goto do_cleanup;
}
printf("listen with %d backlog success!\r\n", backlog);
//阻塞等待客户端的连接请求
connfd = accept(sockfd, (struct sockaddr *)&clientAddr, &clientAddrLen);
if (connfd < 0)
{
    printf("accept failed, %d, %d\r\n", connfd, errno);
    goto do_cleanup;
}
printf("accept success, connfd = %d!\r\n", connfd);
printf("client addr info: host = %s, port = %d\r\n", inet_ntoa(clientAddr.sin_addr),
ntohs(clientAddr.sin_port));
//后续收、发都在表示连接的Socket上进行
retval = recv(connfd, message, sizeof(message), 0);
if (retval < 0)
{
    printf("recv message failed, %ld!\r\n", retval);
    goto do_disconnect;
}
printf("recv message{%s} from client done!\r\n", message);
retval = send(connfd, message, strlen(message), 0);
if (retval <= 0)
{
```

```
        printf("send message failed, %ld!\r\n", retval);
        goto do_disconnect;
    }
    printf("send message{%s} to client done!\r\n", message);
do_disconnect:
    printf("do_disconnect...\r\n");
    closesocket(connfd);
do_cleanup:
    printf("do_cleanup...\r\n");
    closesocket(sockfd);
}
```

(5) 创建方法 connect_wifi，实现 WiFi 连接功能，代码如下：

```
//applications/sample/wifi-iot/app/ohos_50/chapter_05/case46-tcp_server/tcp_server_demo.c
void connect_wifi(void)
{
    WifiErrorCode errCode;
    WifiDeviceConfig apConfig = {};
    int netId = -1;
    strcpy(apConfig.ssid, "Tenda_1F7390");
    strcpy(apConfig.preSharedKey, "nasoftmt");
    apConfig.securityType = WIFI_SEC_TYPE_PSK;
    errCode = EnableWifi();
    errCode = AddDeviceConfig(&apConfig, &netId);
    errCode = ConnectTo(netId);
    printf("ConnectTo(%d): %d\r\n", netId, errCode);
    usleep(1000 * 1000);
    struct netif * iface = netifapi_netif_find("wlan0");
    if (iface)
    {
        err_t ret = netifapi_dhcp_start(iface);
        printf("netifapi_dhcp_start: %d\r\n", ret);
        usleep(2000 * 1000);
        ret = netifapi_netif_common(iface, dhcp_clients_info_show, NULL);
        printf("netifapi_netif_common: %d\r\n", ret);
    }
}
```

(6) 创建 TCP 服务器端任务，实现收发信息功能，代码如下：

```
//applications/sample/wifi-iot/app/ohos_50/chapter_05/case46-tcp_server/tcp_server_demo.c
static void tcp_client_thread(void * arg)
```

```c
{
    (void)arg;
    sleep(1);
    connect_wifi();
    sleep(1);
    printf("tcp client demo start\n");
    unsigned short port = 5001;
    tcp_server(port);
}
static void entry(void)
{
    osThreadAttr_t attr;
    attr.name = "tcp_client_thread";
    attr.attr_bits = 0U;
    attr.cb_mem = NULL;
    attr.cb_size = 0U;
    attr.stack_mem = NULL;
    attr.stack_size = 10240;
    attr.priority = osPriorityNormal;
    if (osThreadNew(tcp_client_thread, NULL, &attr) == NULL)
    {
        printf("[tcp_client_thread] Failed to create tcp_client_thread!\n");
    }
}
APP_FEATURE_INIT(entry);
```

(7) 创建并编写模块,构建脚本 BUILD.gn,代码如下:

```
//applications/sample/wifi-iot/app/ohos_50/chapter_05/case46-tcp_server/BUILD.gn
static_library("ch_05_tcp_server") {
    sources = [
        "tcp_server_demo.c",
    ]
    include_dirs = [
        "//kernel/liteos_m/components/cmsis/2.0",
        "//base/iot_hardware/interfaces/kits/wifiiot_lite",
        "//vendor/hisi/hi3861/hi3861/third_party/lwip_sack/include",
        "//foundation/communication/interfaces/kits/wifi_lite/wifiservice",
    ]
}
```

(8) 将应用模块配置到应用子系统,代码如下:

```
//applications/sample/wifi-iot/app/BUILD.gn
import("//build/lite/config/component/lite_component.gni")
lite_component("app") {
    features = [
```

```
            "ohos_50/chapter_05/case46-tcp_server:ch_05_tcp_server",
    ]
}
```

（9）测试：编译应用模块，将固件烧写到开发板，复位开发板，将开发板与 IPOP 终端工具连接，然后用 NetAssist 网络调试助手创建 TCP 客户端并向 TCP 服务器端发送信息。观察 NetAssist 网络调试助手 TCP 客户端窗口的打印信息，如图 5-14 所示。

图 5-14　TCP 服务器端应用（NetAssist 网络调试助手客户端）的运行效果

观察 IPOP 终端工具窗口的打印信息，如图 5-15 所示。

```
+NOTICE:CONNECTED
netifapi_dhcp_start: 0
server :
        server_id : 192.168.0.1
        mask : 255.255.255.0, 0
        gw : 0.0.0.0
        T0 : 0
        T1 : 0
        T2 : 0
clients <1> :
        mac_idx mac              addr              state    lease    tries
        0       e01131881478     192.168.0.151     1        0        1
netifapi_netif_common: 0
tcp client demo start
bind to port 5001 success!
listen with 1 backlog success!
accept success, connfd = 1!
client addr info: host = 192.168.0.105, port = 61906
recv message{Hello world
} from client done!
send message{Hello world
} to client done!
do_disconnect...
do_cleanup...
```

图 5-15　TCP 服务器端应用（IPOP 终端工具）的运行效果

5.4 物联网

本节将会通过 3 个案例详细介绍关于 MQTT 的物联网开发。

消息队列遥测传输（Message Queuing Telemetry Transport，MQTT）是一个基于客户端-服务器端的消息发布/订阅传输协议。MQTT 协议是基于 TCP/IP 协议的轻量、简单、开放和易于实现的协议，这些特点使它的适用范围非常广泛。在很多情况下，包括受限的环境中，如机器与机器（M2M）通信和物联网（IoT）。其在通过卫星链路通信传感器、偶尔拨号的医疗设备、智能家居及一些小型化设备中被广泛使用。

5.4.1 案例 47：MQTT 第三方库移植

本案例为 MQTT 第三方库移植，介绍将第三方库移植到 OpenHarmony 系统的基本方法及步骤，移植流程如下：

（1）将第三方库源码复制到 OpenHarmony 根目录下的 third_party。

（2）在源码根目录中创建并编写第三方库，构建脚本 BUILD.gn。

（3）在系统模块配置脚本中添加第三方库。

移植步骤如下：

（1）进入网站 https://github.com/eclipse/paho.mqtt.embedded-c，下载 MQTT 源码库，如图 5-16 所示。

图 5-16　MQTT 源码库下载

(2) 在 OpenHarmony 源码的 third_party 目录下创建 pahomqtt 目录，如图 5-17 所示。

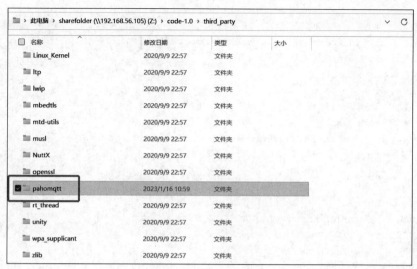

图 5-17　创建 pahomqtt 目录

(3) 将 MQTT 源码解压到 pahomqtt 目录下，如图 5-18 所示。

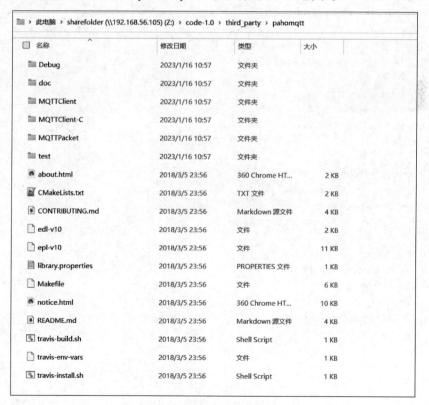

图 5-18　MQTT 源码解压

（4）在pahomqtt目录下创建并编写模块，构建脚本BUILD.gn，代码如下：

```
//third_party/pahomqtt/BUILD.gn
import("//build/lite/config/component/lite_component.gni")
import("//build/lite/ndk/ndk.gni")
config("pahomqtt_config") {
    include_dirs = [
        "MQTTPacket/src",
        "MQTTPacket/samples",
        "//vendor/hisi/hi3861/hi3861/third_party/lwip_sack/include",
        "//kernel/liteos_m/components/cmsis/2.0",
    ]
}
pahomqtt_sources = [
    "MQTTPacket/samples/transport.c",
    "MQTTPacket/src/MQTTConnectClient.c",
    "MQTTPacket/src/MQTTConnectServer.c",
    "MQTTPacket/src/MQTTDeserializePublish.c",
    "MQTTPacket/src/MQTTFormat.c",
    "MQTTPacket/src/MQTTPacket.c",
    "MQTTPacket/src/MQTTSerializePublish.c",
    "MQTTPacket/src/MQTTSubscribeClient.c",
    "MQTTPacket/src/MQTTSubscribeServer.c",
    "MQTTPacket/src/MQTTUnsubscribeClient.c",
    "MQTTPacket/src/MQTTUnsubscribeServer.c",
]
lite_library("pahomqtt_static") {
    target_type = "static_library"
    sources = pahomqtt_sources
    public_configs = [ ":pahomqtt_config" ]
}
lite_library("pahomqtt_shared") {
    target_type = "shared_library"
    sources = pahomqtt_sources
    public_configs = [ ":pahomqtt_config" ]
}
ndk_lib("pahomqtt_ndk") {
    if (board_name != "hi3861v100") {
        lib_extension = ".so"
        deps = [
            ":pahomqtt_shared"
        ]
    } else {
        deps = [
            ":pahomqtt_static"
        ]
```

```
    }
    head_files = [
        "//third_party/pahomqtt"
    ]
}
```

（5）在 vendor/hisi/hi3861/hi3861/BUILD.gn 文件中的 lite_component("sdk") 中将 MQTT 添加为系统模块，代码如下：

```
//vendor/hisi/hi3861/hi3861/BUILD.gn
lite_component("sdk") {
    features = [ ]

    deps = [ "//kernel/liteos_m/components/cmsis",
            "//kernel/liteos_m/components/kal",
            "//third_party/cJSON:cjson_static",
            "//third_party/pahomqtt:pahomqtt_static",
    ]
}
```

（6）在 third_party/pahomqtt/MQTTPacket/samples/transport.c 文件中添加 Socket 相关头文件，代码如下：

```
//third_party/pahomqtt/MQTTPacket/samples/transport.c
#include "lwip/ip_addr.h"
#include "lwip/netifapi.h"
#include "lwip/sockets.h"
```

（7）修改 transport.c 文件中的 transport_sendPacketBuffer 函数，代码如下：

```
//third_party/pahomqtt/MQTTPacket/samples/transport.c
int transport_sendPacketBuffer(int sock, unsigned char* buf, int buflen)
{
    int rc = 0;
    //rc = write(sock, buf, buflen);              //修改前
    rc = send(sock, buf, buflen,0);               //修改后
    return rc;
}
```

（8）修改 transport.c 文件中的 transport_close 函数，代码如下：

```
//third_party/pahomqtt/MQTTPacket/samples/transport.c
int transport_close(int sock)
{
```

```
    int rc;
    rc = shutdown(sock, SHUT_WR);
    rc = recv(sock, NULL, (size_t)0, 0);
    //rc = close(sock);                    //修改前
    rc = lwip_close(sock);                 //修改后
    return rc;
}
```

（9）给所有未使用的函数参数添加一行参数并赋值给本参数的代码，如 buflen = buflen，以此保证编译时不报错。直到编译成功，如图 5-19～图 5-21 所示。

```
../../third_party/pahomqtt/MQTTPacket/src/MQTTConnectClient.c:132:111: error: un
used parameter 'buflen' [-Werror=unused-parameter]
 int MQTTDeserialize_connack(unsigned char* sessionPresent, unsigned char* conna
ck_rc, unsigned char* buf, int buflen)
```

图 5-19　编译报错

```
int MQTTDeserialize_connack(unsigned char* sessionPresent, unsigned char* connack_rc, unsigned char* buf, int buflen)
{
    buflen = buflen;
    MQTTHeader header = {0};
    unsigned char* curdata = buf;
    unsigned char* enddata = NULL;
```

图 5-20　修改代码

```
[common sign][01]=[0x21]
[common sign][30]=[0x2f]
[common sign][31]=[0x82]
[section sign][00]=[0x3]
[section sign][01]=[0x99]
[section sign][30]=[0xe6]
[section sign][31]=[0xec]
[image_id=0x3c78961e][struct_version=0x0]]
[hash_alg=0x0][sign_alg=0x3f][sign_param=0x0]
[section_count=0x1]
[section0_compress=0x0][section0_offset=0x3c0][section0_len=0x5f60]
[section1_compress=0x0][section1_offset=0x0][section1_len=0x0]
--------------output/bin/Hi3861_wifiiot_app_flash_boot_ota.bin image info print e
nd--------------

< ^^^^^^^^^^^^^^^^^^^^^^^^^^^^^^^^^^^^^^^^^^^^^^^^^^^^^^^^^^^^^^^ >
                          BUILD SUCCESS
< ^^^^^^^^^^^^^^^^^^^^^^^^^^^^^^^^^^^^^^^^^^^^^^^^^^^^^^^^^^^^^^^ >

See build log from: /home/gs/sharefolder/code-1.0/vendor/hisi/hi3861/hi3861/buil
d/build_tmp/logs/build_kernel.log
[212/212] STAMP obj/vendor/hisi/hi3861/hi3861/run_wifiiot_scons.stamp
ohos wifiiot build success!
gs@gs-ohos:~/sharefolder/code-1.0$ python build.py wifiiot
```

图 5-21　编译成功

5.4.2 案例48：MQTT协议应用

本案例为MQTT协议应用，讲解MQTT客户端工具与服务器端的基本应用场景。MQTT的基本操作流程如下：

(1) 安装并启动MQTT服务程序。
(2) 安装并运行MQTT客户端程序。
(3) 客户端程序连接MQTT服务。
(4) 客户端订阅主题。
(5) 客户端发布主题信息。

实现步骤如下：

(1) 进入网站https://www.emqx.io/cn/downloads#broker，下载MQTT服务器端，如图5-22所示。

图5-22 下载MQTT服务器端

(2) 解压MQTT服务器端，如图5-23所示。
(3) 通过命令emqx start启动MQTT服务器，如图5-24所示。
(4) 进入网站http://127.0.0.1:18083/，进入服务器后台，如图5-25所示。
(5) 登录MQTT服务器后台(默认用户名为admin，默认密码为public)，如图5-26所示。
(6) 更改密码，如图5-27所示。
(7) 进入网站https://repo.eclipse.org/content/repositories/pahoreleases/org/eclipse/paho/org.eclipse.paho.ui.app/，下载MQTT客户端，如图5-28所示。
(8) 解压完成后，双击运行paho.exe。

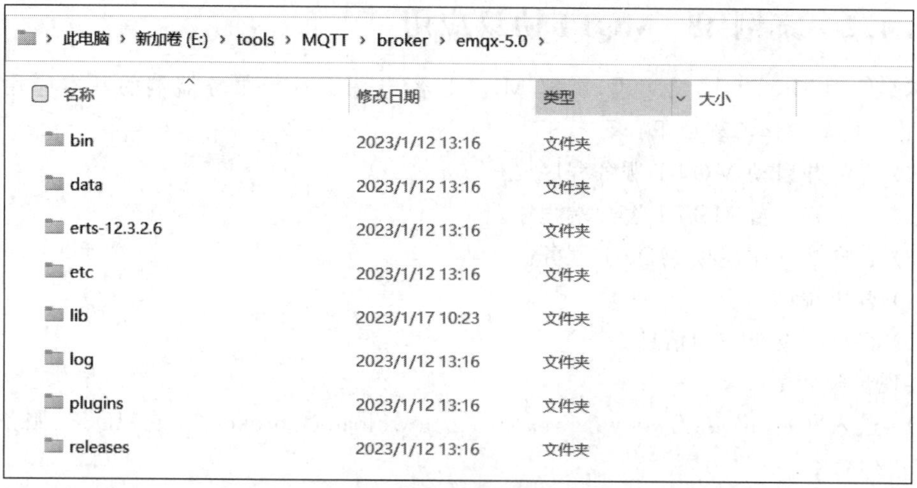

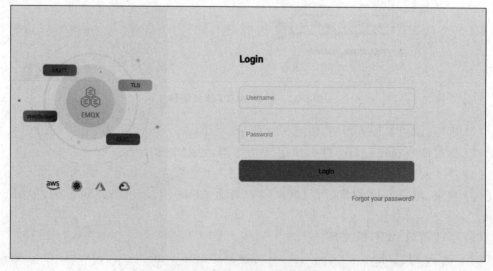

图 5-23　解压 MQTT 服务器端

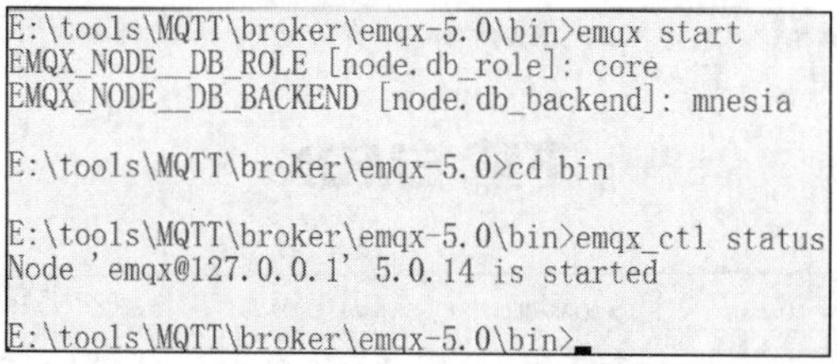

图 5-24　启动 MQTT 服务器

图 5-25　进入 MQTT 服务器后台

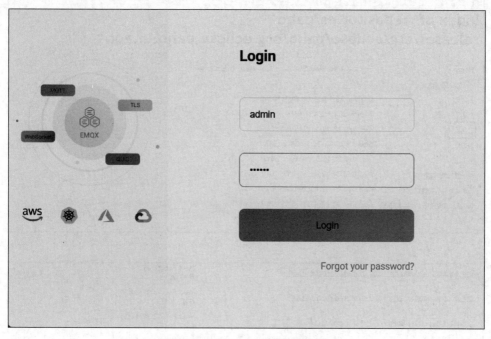

图 5-26 登录 MQTT 服务器后台

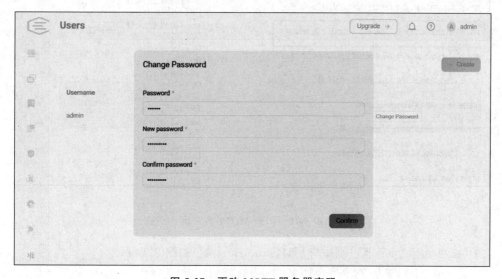

图 5-27 更改 MQTT 服务器密码

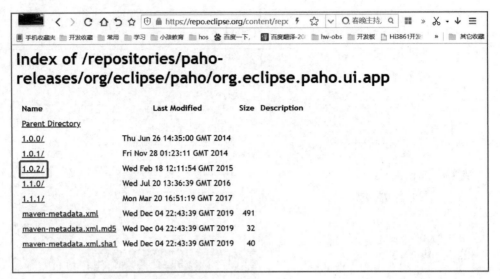

图 5-28 下载 MQTT 客户端

（9）在两个 MQTT 客户端中分别订阅 music 主题，如图 5-29 所示。

（10）查看 MQTT 服务器后台信息，如图 5-30 所示。

（11）在一个 MQTT 客户端中发布主题为 music 的信息，如图 5-31 所示。

（12）测试效果如图 5-32 所示。

第5章 OpenHarmony轻量系统物联网开发

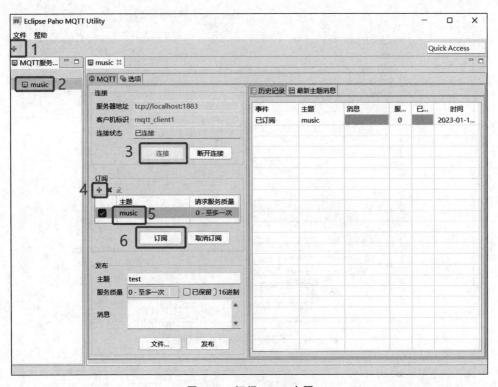

图 5-29 订阅 music 主题

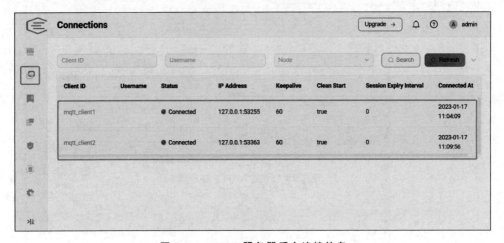

图 5-30 MQTT 服务器后台连接信息

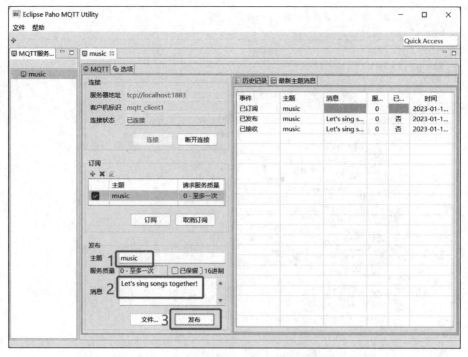

图 5-31　在 MQTT 客户端发布信息

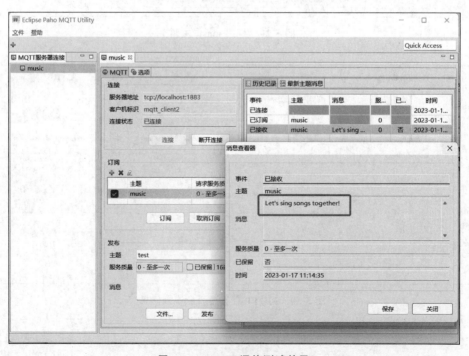

图 5-32　MQTT 通信测试效果

5.4.3 案例49:MQTT物联网应用开发

本案例为MQTT物联网应用开发,实现MQTT主题订阅和发布功能。MQTT应用开发的流程如下:

(1) 配置MQTT服务器端的IP地址和端口。
(2) 连接MQTT服务器。
(3) 订阅/发布主题。
(4) 取消订阅。
(5) 断开服务器连接。

开发步骤如下:
(1) 在本章目录chapter_05中创建应用模块工程目录case49-mqtt_client。
(2) 在应用模块工程目录case49-mqtt_client中创建MQTT工具文件mqtt_util.c。
(3) 编写MQTT工具文件,代码如下:

```c
//applications/sample/wifi-iot/app/ohos_50/chapter_05/case49-mqtt_client/mqtt_util.c
#include <stdio.h>
#include <unistd.h>
#include "ohos_init.h"
#include "cmsis_os2.h"
#include <unistd.h>
#include "hi_wifi_api.h"
#include "lwip/ip_addr.h"
#include "lwip/netifapi.h"
#include "lwip/sockets.h"
#include "MQTTPacket.h"
#include "transport.h"
int toStop = 0;

int mqtt_connect(void)
{
    MQTTPacket_connectData data = MQTTPacket_connectData_initializer;
    int rc = 0;
    int mysock = 0;
    unsigned char buf[200];
    int buflen = sizeof(buf);
    int msgid = 1;
    MQTTString topicString = MQTTString_initializer;
    int req_qos = 0;
    char payload[200] = "publish....";
    int payloadlen = strlen(payload);
    int len = 0;
    char *host = "192.168.0.105";              //MQTT服务器的IP地址
```

```c
    //char * host = "121.36.35.193";                    //MQTT服务器的IP地址
    int port = 1883;
    mysock = transport_open(host, port);
    if (mysock < 0)
        return mysock;
    printf("Sending to hostname %s port %d\n", host, port);
    data.clientID.cstring = "Hi3861_gs";                //修改成自己开发板的名称
    data.keepAliveInterval = 20;
    data.cleansession = 1;
    data.username.cstring = "Hi3861_gs";
    data.password.cstring = "password";
    len = MQTTSerialize_connect(buf, buflen, &data);
    rc = transport_sendPacketBuffer(mysock, buf, len);
    /* 等待连接 */
    if (MQTTPacket_read(buf, buflen, transport_getdata) == CONNACK)
    {
        unsigned char sessionPresent, connack_rc;
        if (MQTTDeserialize_connack(&sessionPresent, &connack_rc, buf, buflen) != 1 || connack_rc != 0)
        {
            printf("Unable to connect, return code %d\n", connack_rc);
            goto exit;
        }
    }
    else
        goto exit;
    /* 订阅 */
    topicString.cstring = "test";                       //订阅主题
    len = MQTTSerialize_subscribe(buf, buflen, 0, msgid, 1, &topicString, &req_qos);
    rc = transport_sendPacketBuffer(mysock, buf, len);
    if (MQTTPacket_read(buf, buflen, transport_getdata) == SUBACK) /* wait for suback */
    {
        unsigned short submsgid;
        int subcount;
        int granted_qos;
        rc = MQTTDeserialize_suback(&submsgid, 1, &subcount, &granted_qos, buf, buflen);
        if (granted_qos != 0)
        {
            printf("granted qos != 0, %d\n", granted_qos);
            goto exit;
        }
    }
    else
        goto exit;
    /* 循环获取已订阅主题的消息 */
    while (!toStop)
```

```
{
    /* transport_getdata() has a built-in 1 second timeout,
    your mileage will vary */
    if (MQTTPacket_read(buf, buflen, transport_getdata) == PUBLISH)
    {
        unsigned char dup;
        int qos;
        unsigned char retained;
        unsigned short msgid;
        int payloadlen_in;
        unsigned char * payload_in;
        int rc;
        MQTTString receivedTopic;
        rc = MQTTDeserialize_publish(&dup, &qos, &retained, &msgid, &receivedTopic,
                            &payload_in, &payloadlen_in, buf, buflen);
        printf("message arrived %d, %s\n", payloadlen_in, payload_in);
        rc = rc;
    }
    topicString.cstring = "publish_topic";           //设置发布主题
    if (buflen <= 0)
    {
        strcpy(payload, "send from hi3861");
        payloadlen = strlen(payload);
    }
    len = MQTTSerialize_publish(buf, buflen, 0, 0, 0, 0, topicString, (unsigned char *)
payload, payloadlen);
    rc = transport_sendPacketBuffer(mysock, buf, len);
    usleep(100000);
    }
    printf("disconnecting\n");
    len = MQTTSerialize_disconnect(buf, buflen);
    rc = transport_sendPacketBuffer(mysock, buf, len);
exit:
    transport_close(mysock);
    rc = rc;
    return 0;
}
```

(4) 创建并编写 MQTT 工具文件对应的头文件 mqtt_util.h,代码如下：

```
//applications/sample/wifi-iot/app/ohos_50/chapter_05/case49-mqtt_client/mqtt_util.h
#ifndef MQTT_UTIL
#define MQTT_UTIL
void mqtt_connect(void);
#endif
```

(5) 创建并编写应用模块源码文件 mqtt_client_demo.c, 代码如下：

```c
//applications/sample/wifi-iot/app/ohos_50/chapter_05/case49-mqtt_client/mqtt_client_demo.c
#include <stdio.h>
#include <string.h>
#include <unistd.h>
#include "ohos_init.h"
#include "cmsis_os2.h"
#include "wifi_device.h"
#include "lwip/netifapi.h"
#include "lwip/api_shell.h"
#include "mqtt_util.h"
void connect_wifi(void)
{
    WifiErrorCode errCode;
    WifiDeviceConfig apConfig = {};
    int netId = -1;
    strcpy(apConfig.ssid, "Tenda_1F7390");
    strcpy(apConfig.preSharedKey, "nasoftmt");
    apConfig.securityType = WIFI_SEC_TYPE_PSK;
    errCode = EnableWifi();
    errCode = AddDeviceConfig(&apConfig, &netId);
    errCode = ConnectTo(netId);
    printf("ConnectTo(%d): %d\r\n", netId, errCode);
    usleep(1000 * 1000);
    struct netif *iface = netifapi_netif_find("wlan0");
    if (iface)
    {
        err_t ret = netifapi_dhcp_start(iface);
        printf("netifapi_dhcp_start: %d\r\n", ret);
        usleep(2000 * 1000);
        ret = netifapi_netif_common(iface, dhcp_clients_info_show, NULL);
        printf("netifapi_netif_common: %d\r\n", ret);
    }
}
static void mqtt_client_thread(void *arg)
{
    (void)arg;
    sleep(1);
    connect_wifi();
    sleep(1);
    printf("mqtt client demo start\n");
    mqtt_connect();
}
static void entry(void)
```

```
{
    osThreadAttr_t attr;
    attr.name = "mqtt_client_thread";
    attr.attr_bits = 0U;
    attr.cb_mem = NULL;
    attr.cb_size = 0U;
    attr.stack_mem = NULL;
    attr.stack_size = 10240;
    attr.priority = osPriorityNormal;
    if (osThreadNew(mqtt_client_thread, NULL, &attr) == NULL)
    {
        printf("[mqtt_client_thread] Failed to create mqtt_client_thread!\n");
    }
}
APP_FEATURE_INIT(entry);
```

(6) 创建并编写模块,构建脚本 BUILD.gn,代码如下:

```
//applications/sample/wifi-iot/app/ohos_50/chapter_05/case49-mqtt_client/BUILD.gn
static_library("ch_05_mqtt_client") {
    sources = [
        "mqtt_client_demo.c",
        "mqtt_util.c",
    ]
    include_dirs = [
        "//third_party/pahomqtt/MQTTPacket/src",
        "//third_party/pahomqtt/MQTTPacket/samples",
        "//kernel/liteos_m/components/cmsis/2.0",
        "//base/iot_hardware/interfaces/kits/wifiiot_lite",
        "//vendor/hisi/hi3861/hi3861/third_party/lwip_sack/include",
        "//foundation/communication/interfaces/kits/wifi_lite/wifiservice",
    ]
}
```

(7) 将应用模块配置到应用子系统,代码如下:

```
//applications/sample/wifi-iot/app/BUILD.gn
import("//build/lite/config/component/lite_component.gni")
lite_component("app") {
    features = [
        "ohos_50/chapter_05/case49-mqtt_client:ch_05_mqtt_client",
    ]
}
```

（8）测试：编译应用模块，将固件烧写到开发板，复位开发板，将开发板与 IPOP 终端工具连接。使用 MQTT 客户端订阅主题 publish_topic，如图 5-33 所示。

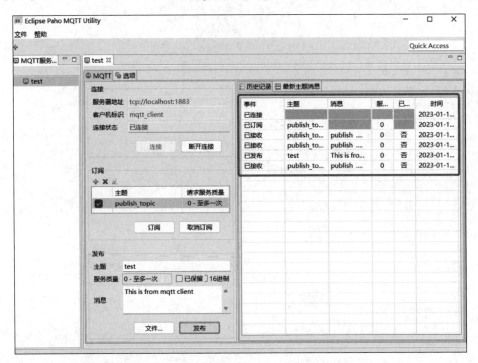

图 5-33　MQTT 物联网应用开发（MQTT 客户端）的运行效果

观察 IPOP 终端工具窗口的打印信息，如图 5-34 所示。

```
ConnectTo(0): 0
+NOTICE:SCANFINISH
+NOTICE:CONNECTED
netifapi_dhcp_start: 0
server :
        server_id : 192.168.0.1
        mask : 255.255.255.0, 0
        gw : 0.0.0.0
        T0 : 0
        T1 : 0
        T2 : 0
clients <1> :
        mac_idx mac              addr            state   lease   tries
        0       181131e4159e     192.168.0.117   1       0       1
netifapi_netif_common: 0
mqtt_client_demo_start
Sending to hostname 192.168.0.105 port 1883
message arrived 24,This is from mqtt clientgs
```

图 5-34　MQTT 物联网应用开发（IPOP 终端工具）的运行效果

注意：如果在运行时连接不上 MQTT 服务器端，则应在计算机设置中关闭防火墙，或开放所需端口。

第 6 章 综合案例：遥控小车

6.1 案例介绍

6.1.1 案例架构介绍

本案例通过 HarmonyOS 端 App 实现远程操控 OpenHarmony 遥控小车的功能。本案例整体分为 HarmonyOS 手表端、服务器端和 OpenHarmony 开发板端。各端的功能具体如下：

（1）HarmonyOS 手表端：运行 Java 模板遥控小车 App，App 中的 MQTT Client 线程与服务器端的 MQTT 服务程序相连，用户通过操作 App 的 UI 完成小车指令的相关主题消息发布。

（2）服务器端：运行 MQTT 服务程序，实现提供 MQTT 主题订阅、发布、控制台服务功能。

（3）OpenHarmony 开发板端：运行遥控小车应用 App，App 中包含实现指令接收及小车移动控制功能。

6.1.2 技术架构图

案例中 HarmonyOS 手表端、服务器端和 OpenHarmony 开发板端的技术架构如图 6-1 所示。

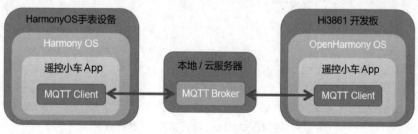

图 6-1 技术架构

6.1.3 运行效果

小车端的运行效果如图 6-2 所示。

图 6-2 遥控小车（小车端）的运行效果

小车端 IPOP 终端工具打印的信息如图 6-3 所示。

```
remote_control_car running
car_working() start
car_mode_control_func start
ConnectTo(0): 0
+NOTICE:SCANFINISH
+NOTICE:CONNECTED
netifapi_dhcp_start: 0
server :
        server_id : 192.168.0.1
        mask : 255.255.255.0, 0
        gw : 0.0.0.0
        T0 : 0
        T1 : 0
        T2 : 0
clients <1> :
        mac_idx mac              addr              state    lease    tries
        0       b4c9b9af689e     192.168.0.104     1        0        1
netifapi_netif_common: 0
mqtt client demo start
Sending to hostname 192.168.0.103 port 1883
message arrived 1.1s
remote_control_car_cmd:1
message arrived 1.0s
remote_control_car_cmd:0
message arrived 1.2s
remote_control_car_cmd:2
message arrived 1.0s
remote_control_car_cmd:0
```

图 6-3 遥控小车（IPOP 终端工具）的运行效果

手表端的运行效果如图 6-4 所示。

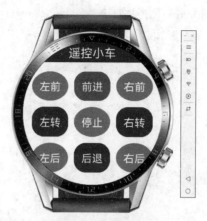

图 6-4　遥控小车（手表端）的运行效果

服务器端后台的连接效果如图 6-5 所示。

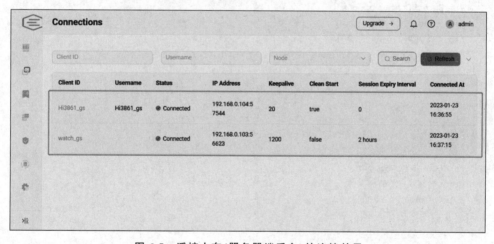

图 6-5　遥控小车（服务器端后台）的连接效果

6.2　OpenHarmony 开发板端功能实现

在 OpenHarmony 开发板端创建 OpenHarmony 遥控小车应用 App，实现指令接收及小车移动控制功能。App 中包含了 MQTT 通信模块、小车指令执行模块和主模块。

6.2.1　MQTT 通信模块功能实现

MQTT 通信模块实现 WiFi 联网和 MQTT 通信功能，具体的实现步骤如下：
（1）在 ohos_50 中创建本章源码目录 chapter_06。

（2）在 chapter_06 中创建遥控小车应用 App 工程目录 case50-remote_control_car。

（3）在 case50-remote_control_car 中创建并编写文件 wifi_util.c 实现 WiFi 联网功能，代码如下：

```c
//applications/sample/wifi-iot/app/ohos_50/chapter_06/case50-remote_control_car/wifi_util.c
#include <stdio.h>
#include <string.h>
#include <unistd.h>
#include "ohos_init.h"
#include "cmsis_os2.h"
#include "wifi_device.h"
#include "lwip/netifapi.h"
#include "lwip/api_shell.h"
void connect_wifi(void)
{
    WifiErrorCode errCode;
    WifiDeviceConfig apConfig = {};
    int netId = -1;
    //配置将要连接的 AP 属性
    strcpy(apConfig.ssid, "Tenda_1F7390");                  //网络名称
    strcpy(apConfig.preSharedKey, "nasoftmt");              //密码
    apConfig.securityType = WIFI_SEC_TYPE_PSK;              //加密类型
    errCode = EnableWifi();                                 //启用 STA 模式
    errCode = AddDeviceConfig(&apConfig, &netId);           //配置热点信息,生成网络 ID
    printf("AddDeviceConfig: %d\r\n", errCode);

    errCode = ConnectTo(netId);                             //连接到指定的网络
    printf("ConnectTo(%d): %d\r\n", netId, errCode);
    usleep(3000 * 1000);                                    //等待连接
    //获取网络接口,用于 IP 操作
    struct netif *iface = netifapi_netif_find("wlan0");
    if (iface)
    {
        //启动 DHCP 客户端,获取 IP 地址
        err_t ret = netifapi_dhcp_start(iface);
        printf("netifapi_dhcp_start: %d\r\n", ret);
        usleep(2000 * 1000);
        //netifapi_netif_common 用于以线程安全的方式调用所有与 netif 相关的 API
        //dhcp_clients_info_show 为 Shell API,展示 dhcp 客户端信息
        ret = netifapi_netif_common(iface, dhcp_clients_info_show, NULL);
        printf("netifapi_netif_common: %d\r\n", ret);
    }
}
```

(4) 创建并编写头文件 wifi_util.h，完成函数 connect_wifi 的声明，代码如下：

```
//applications/sample/wifi-iot/app/ohos_50/chapter_06/case50-remote_control_car/wifi_util.h
#ifndef WIFI_UTIL
#define WIFI_UTIL
void connect_wifi(void);
#endif
```

(5) 在 case50-remote_control_car 中创建并编写工具文件 mqtt_util.c，实现 MQTT 服务器连接、小车控制指令主题订阅和指令的解析并将指令更新到外部变量 remote_control_car_stat 中的功能，代码如下：

```c
//applications/sample/wifi-iot/app/ohos_50/chapter_06/case50-remote_control_car/mqtt_util.c
#include <stdio.h>
#include <unistd.h>
#include "ohos_init.h"
#include "cmsis_os2.h"
#include <unistd.h>
#include "hi_wifi_api.h"
#include "lwip/ip_addr.h"
#include "lwip/netifapi.h"
#include "lwip/sockets.h"
#include "MQTTPacket.h"
#include "transport.h"
#include "car_motion.h"
int mqtt_connect(void)
{
    MQTTPacket_connectData data = MQTTPacket_connectData_initializer;
    int rc = 0;
    int mysock = 0;
    unsigned char buf[200];
    int buflen = sizeof(buf);
    int msgid = 1;
    MQTTString topicString = MQTTString_initializer;
    int req_qos = 0;
    int len = 0;
    char *host = "192.168.0.103";          //MQTT 服务器的 IP 地址
    int port = 1883;                        //MQTT 服务器端端口
    mysock = transport_open(host, port);
    if (mysock < 0)
        return mysock;
    printf("Sending to hostname %s port %d\n", host, port);
```

```c
        data.clientID.cstring = "Hi3861_gs";                //修改成自己开发板的名称
        data.keepAliveInterval = 20;
        data.cleansession = 1;
        data.username.cstring = "Hi3861_gs";
        data.password.cstring = "password";
        len = MQTTSerialize_connect(buf, buflen, &data);
        rc = transport_sendPacketBuffer(mysock, buf, len);
        /* 等待连接 */
        if (MQTTPacket_read(buf, buflen, transport_getdata) == CONNACK)
        {
            unsigned char sessionPresent, connack_rc;

            if (MQTTDeserialize_connack(&sessionPresent, &connack_rc, buf, buflen) != 1 || connack_rc != 0)
            {
                printf("Unable to connect, return code %d\n", connack_rc);
                goto exit;
            }
        }
        else
            goto exit;
        /* 订阅 */
        topicString.cstring = "res";                         //订阅主题
        len = MQTTSerialize_subscribe(buf, buflen, 0, msgid, 1, &topicString, &req_qos);
        rc = transport_sendPacketBuffer(mysock, buf, len);
        if (MQTTPacket_read(buf, buflen, transport_getdata) == SUBACK)  /* wait for suback */
        {
            unsigned short submsgid;
            int subcount;
            int granted_qos;
            rc = MQTTDeserialize_suback(&submsgid, 1, &subcount, &granted_qos, buf, buflen);
            if (granted_qos != 0)
            {
                printf("granted qos != 0, %d\n", granted_qos);
                goto exit;
            }
        }
        else
            goto exit;
        /* 循环获取已订阅主题的消息 */
        while (1)
        {
            if (MQTTPacket_read(buf, buflen, transport_getdata) == PUBLISH)
            {
                unsigned char dup;
                int qos;
```

```
            unsigned char retained;
            unsigned short msgid;
            int payloadlen_in;
            unsigned char * payload_in;
            int rc;
            MQTTString receivedTopic;
            rc = MQTTDeserialize_publish(&dup, &qos, &retained, &msgid, &receivedTopic,
                             &payload_in, &payloadlen_in, buf, buflen);
            printf("message arrived %d, %s\n", payloadlen_in, payload_in);
            if (payloadlen_in > 0 && payloadlen_in < 30)
            {
                //解析指令并保存到外部变量 remote_control_car_stat 中
                remote_control_car_stat = payload_in[0] - '0';
                printf("remote_control_car_cmd: %d\r\n", payload_in[0] - '0');
            }
            rc = rc;
        }
        usleep(100000);
    }
    printf("disconnecting\n");
    len = MQTTSerialize_disconnect(buf, buflen);
    rc = transport_sendPacketBuffer(mysock, buf, len);
exit:
    transport_close(mysock);
    rc = rc;
    printf("go to exit\n");
    return 0;
}
```

(6) 创建并编写头文件 mqtt_util.h，完成函数 mqtt_connect 的声明，代码如下：

```
//applications/sample/wifi-iot/app/ohos_50/chapter_06/case50-remote_control_car/mqtt_util.h
#ifndef MQTT_UTIL
#define MQTT_UTIL
void mqtt_connect(void);
#endif
```

(7) 创建并编写 MQTT 通信模块源码文件 mqtt_task.c，实现连接网络与 MQTT 服务器功能，代码如下：

```
//applications/sample/wifi-iot/app/ohos_50/chapter_06/case50-remote_control_car/mqtt_task.c
#include <stdio.h>
#include <string.h>
```

```c
#include <unistd.h>
#include "ohos_init.h"
#include "cmsis_os2.h"
#include "wifi_device.h"
#include "lwip/netifapi.h"
#include "lwip/api_shell.h"
#include "mqtt_util.h"
#include "wifi_util.h"
static void mqtt_client_thread(void *arg)
{
    (void)arg;
    //连接网络
    connect_wifi();
    sleep(1);

    printf("mqtt client demo start\n");
    //调用 MQTT 功能函数
    mqtt_connect();
}
void mqtt_task_entry(void)
{
    osThreadAttr_t attr;
    attr.name = "mqtt_client_thread";
    attr.attr_bits = 0U;
    attr.cb_mem = NULL;
    attr.cb_size = 0U;
    attr.stack_mem = NULL;
    attr.stack_size = 10240;
    attr.priority = osPriorityNormal;
    if (osThreadNew(mqtt_client_thread, NULL, &attr) == NULL)
    {
        printf("[mqtt_client_thread] Failed to create mqtt_client_thread!\n");
    }
}
```

(8) 创建并编写头文件 mqtt_task.h,声明模块入口函数 mqtt_task_entry,代码如下:

```c
//applications/sample/wifi-iot/app/ohos_50/chapter_06/case50-remote_control_car/mqtt_task.h
#ifndef MQTT_TASK
#define MQTT_TASK
void mqtt_task_entry(void);
#endif
```

6.2.2 小车指令执行模块功能实现

(1) 在 case50-remote_control_car 中创建小车指令执行模块源码文件 car_motion.c。

(2) 在 car_motion.c 文件中实现 GPIO 引脚的初始化、指令(小车的指令包括前进、后退、停止、左转、右转、左前、右前、左后和右后)功能函数和模块入口函数,代码如下:

```c
//applications/sample/wifi-iot/app/ohos_50/chapter_06/case50-remote_control_car/car_motion.c
#include <stdio.h>
#include <stdlib.h>
#include <unistd.h>
#include <memory.h>
#include "ohos_init.h"
#include "cmsis_os2.h"
#include "hi_wifi_api.h"
#include "wifiiot_gpio.h"
#include "wifiiot_gpio_ex.h"
#include "wifiiot_pwm.h"
#include "hi_io.h"
#include "hi_time.h"
#include "hi_adc.h"
#include "wifiiot_errno.h"
#include "hi_stdlib.h"
#include "wifiiot_watchdog.h"
#include "mqtt_util.h"
#include "car_motion.h"
//定义全局变量
int remote_control_car_stat;
void gpio_init(void)
{
    GpioInit();
    //引脚复用
    IoSetFunc(WIFI_IOT_IO_NAME_GPIO_0, WIFI_IOT_IO_FUNC_GPIO_0_GPIO);
    IoSetFunc(WIFI_IOT_IO_NAME_GPIO_1, WIFI_IOT_IO_FUNC_GPIO_1_GPIO);
    IoSetFunc(WIFI_IOT_IO_NAME_GPIO_9, WIFI_IOT_IO_FUNC_GPIO_9_GPIO);
    IoSetFunc(WIFI_IOT_IO_NAME_GPIO_10, WIFI_IOT_IO_FUNC_GPIO_10_GPIO);
    //初始化 GPIO
    GpioSetDir(WIFI_IOT_IO_NAME_GPIO_0,WIFI_IOT_GPIO_DIR_OUT);
    GpioSetDir(WIFI_IOT_IO_NAME_GPIO_1,WIFI_IOT_GPIO_DIR_OUT);
    GpioSetDir(WIFI_IOT_IO_NAME_GPIO_9,WIFI_IOT_GPIO_DIR_OUT);
    GpioSetDir(WIFI_IOT_IO_NAME_GPIO_10,WIFI_IOT_GPIO_DIR_OUT);
}
//停止
void car_stop(void)
{
```

```c
    //先停止 GPIO
    GpioSetOutputVal(WIFI_IOT_IO_NAME_GPIO_0,WIFI_IOT_GPIO_VALUE0);
    GpioSetOutputVal(WIFI_IOT_IO_NAME_GPIO_1,WIFI_IOT_GPIO_VALUE0);
    GpioSetOutputVal(WIFI_IOT_IO_NAME_GPIO_9,WIFI_IOT_GPIO_VALUE0);
    GpioSetOutputVal(WIFI_IOT_IO_NAME_GPIO_10,WIFI_IOT_GPIO_VALUE0);
    remote_control_car_stat = CAR_STATUS_STOP;
}
//前进
void car_forward(void)
{
    car_stop();
    //启动 A 路 PWM
    GpioSetOutputVal(WIFI_IOT_IO_NAME_GPIO_0,WIFI_IOT_GPIO_VALUE1);
    GpioSetOutputVal(WIFI_IOT_IO_NAME_GPIO_10,WIFI_IOT_GPIO_VALUE1);
    remote_control_car_stat = CAR_STATUS_FORWARD;
}
//后退
void car_backward(void)
{
    car_stop();
    //启动 B 路 PWM
    GpioSetOutputVal(WIFI_IOT_IO_NAME_GPIO_1,WIFI_IOT_GPIO_VALUE1);
    GpioSetOutputVal(WIFI_IOT_IO_NAME_GPIO_9,WIFI_IOT_GPIO_VALUE1);
    remote_control_car_stat = CAR_STATUS_BACKWARD;
}
//左转
void car_left(void)
{
    car_stop();
    //启动左轮 A 路 PWM
    GpioSetOutputVal(WIFI_IOT_IO_NAME_GPIO_0,WIFI_IOT_GPIO_VALUE1);
    remote_control_car_stat = CAR_STATUS_LEFT;
}
//左前
void car_left_front(void)
{
    car_stop();
    //启动左轮 A 路 PWM
    GpioSetOutputVal(WIFI_IOT_IO_NAME_GPIO_0,WIFI_IOT_GPIO_VALUE1);
    remote_control_car_stat = CAR_STATUS_LEFT_FRONT;
    usleep(100000);
    car_stop();
}
//左后
void car_left_back(void)
{
```

```c
    car_stop();
    //启动右轮B路PWM
    GpioSetOutputVal(WIFI_IOT_IO_NAME_GPIO_1,WIFI_IOT_GPIO_VALUE1);
    remote_control_car_stat = CAR_STATUS_LEFT_FRONT;
    usleep(70000);
    car_stop();
}
//右转
void car_right(void)
{
    car_stop();
    //启动右轮A路PWM
    GpioSetOutputVal(WIFI_IOT_IO_NAME_GPIO_10,WIFI_IOT_GPIO_VALUE1);

    remote_control_car_stat = CAR_STATUS_RIGHT;
}
//右前
void car_right_front(void)
{
    car_stop();
    //启动右轮A路PWM
    GpioSetOutputVal(WIFI_IOT_IO_NAME_GPIO_10,WIFI_IOT_GPIO_VALUE1);
    remote_control_car_stat = CAR_STATUS_RIGHT_FRONT;
    usleep(70000);
    car_stop();
}
//右后
void car_right_back(void)
{
    car_stop();
    //启动右轮B路PWM
    GpioSetOutputVal(WIFI_IOT_IO_NAME_GPIO_9,WIFI_IOT_GPIO_VALUE1);
    remote_control_car_stat = CAR_STATUS_RIGHT_FRONT;
    usleep(70000);
    car_stop();
}
void car_mode_control_func(void)
{
    printf("car_mode_control_func start\r\n");
    while (1)
    {
        WatchDogDisable();
        switch (remote_control_car_stat)
        {
            case CAR_STATUS_STOP:
                car_stop();
```

```c
                break;
            case CAR_STATUS_FORWARD:
                car_forward();
                break;
            case CAR_STATUS_LEFT:
                car_left();
                break;
            case CAR_STATUS_LEFT_FRONT:
                car_left_front();
                break;
            case CAR_STATUS_LEFT_BACK:
                car_left_back();
                break;
            case CAR_STATUS_RIGHT:
                car_right();
                break;
            case CAR_STATUS_RIGHT_FRONT:
                car_right_front();
                break;
            case CAR_STATUS_RIGHT_BACK:
                car_right_back();
                break;
            case CAR_STATUS_BACKWARD:
                car_backward();
                break;
            default:
                car_stop();
                break;
        }
        usleep(500000);
    }
}
void car_working(void)
{
    remote_control_car_stat = CAR_STATUS_STOP;
    gpio_init();
    printf("car_working() start\r\n");
    car_mode_control_func();
}
void rescue_car(void)
{
    osThreadAttr_t attr;
    attr.name = "remote_control_car";
    attr.attr_bits = 0U;
    attr.cb_mem = NULL;
    attr.cb_size = 0U;
```

```
    attr.stack_mem = NULL;
    attr.stack_size = 4096;
    attr.priority = 36;
    if (osThreadNew((osThreadFunc_t)car_working, NULL, &attr) == NULL)
    {
        printf("[remote_control_car] Failed to create car_working!\n");
    }
}
```

(3) 创建并编写头文件 car_motion.h，声明小车指令枚举和小车指令外部全局变量 remote_control_car_stat，代码如下：

```
//applications/sample/wifi-iot/app/ohos_50/chapter_06/case50-remote_control_car/car_motion.h
#ifndef __CAR_MOTION__
#define __CAR_MOTION__
typedef enum
{
    /*停止*/
    CAR_STATUS_STOP = 0,
    /*前进*/
    CAR_STATUS_FORWARD,
    /*后退*/
    CAR_STATUS_BACKWARD,

    /*左转*/
    CAR_STATUS_LEFT,
    /*左前*/
    CAR_STATUS_LEFT_FRONT,
    /*左退*/
    CAR_STATUS_LEFT_BACK,
    /*右转*/
    CAR_STATUS_RIGHT,
    /*右前*/
    CAR_STATUS_RIGHT_FRONT,
    /*右退*/
    CAR_STATUS_RIGHT_BACK,
    /** Maximum value */
    CAR_STATUS_MAX
} CarStatus;
void car_working(void);
//声明为外部全局变量
extern int remote_control_car_stat;
#endif
```

6.2.3　主模块功能实现及测试

（1）创建并编写主模块源码文件 main.c，实现调用 MQTT 通信和小车指令执行模块，代码如下：

```
//applications/sample/wifi-iot/app/ohos_50/chapter_06/case50-remote_control_car/main.c
#include <stdio.h>
#include <unistd.h>
#include "ohos_init.h"
#include "cmsis_os2.h"
#include "mqtt_task.h"
#include "car_motion.h"
void main_thread(void *argv)
{
    (void)argv;
    sleep(1);
    printf("remote_control_car running\n");
    mqtt_task_entry();
    car_working();
}
void MainEntry(void)
{
    osThreadAttr_t attr;
    attr.name = "main_thread";
    attr.attr_bits = 0U;
    attr.cb_mem = NULL;
    attr.cb_size = 0U;
    attr.stack_mem = NULL;
    attr.stack_size = 4096;
    attr.priority = 36;
    if (osThreadNew((osThreadFunc_t)main_thread, NULL, &attr) == NULL)
    {
        printf("[MainEntry] Failed to create main_thread!\n");
    }
}
APP_FEATURE_INIT(MainEntry);
```

（2）创建并编写小车 App 模块，构建脚本 BUILD.gn，配置 App 模块名称、源码文件和头文件目录，代码如下：

```
//applications/sample/wifi-iot/app/ohos_50/chapter_06/case50-remote_control_car/BUILD.gn
static_library("ch_06_remote_control_car") {
    sources = [
```

```
    "mqtt_task.c",
    "mqtt_util.c",
    "WIFI_util.c",
    "car_motion.c",
    "main.c"
  ]
  include_dirs = [
    "//third_party/pahomqtt/MQTTPacket/src",
    "//third_party/pahomqtt/MQTTPacket/samples",
    "//kernel/liteos_m/components/cmsis/2.0",
    "//base/iot_hardware/interfaces/kits/wifiiot_lite",
    "//vendor/hisi/hi3861/hi3861/third_party/lwip_sack/include",
    "//foundation/communication/interfaces/kits/wifi_lite/wifiservice",
  ]
}
```

(3) 将 App 模块 ch_06_remote_control_car 配置到应用子系统中,代码如下:

```
//applications/sample/wifi-iot/app/BUILD.gn
import("//build/lite/config/component/lite_component.gni")
lite_component("app") {
    features = [
        "ohos_50/chapter_06/case50-remote_control_car:ch_06_remote_control_car",,
    ]
}
```

(4) 组装 OpenHarmony 小车开发板,如图 6-6 所示。

图 6-6　组装 OpenHarmony 小车开发板

（5）测试：编译应用模块，将固件烧写到开发板，复位开发板，将开发板与 IPOP 终端工具连接。使用 MQTT 客户端发布主题 res 的信息，参考头文件 car_motion.h 中的小车指令枚举进行测试，如图 6-7 所示。

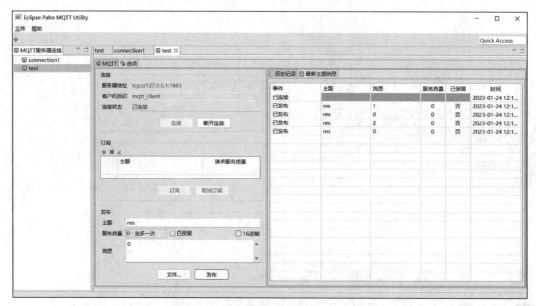

图 6-7　小车端（MQTT 客户端）的运行效果

观察小车运行状态和 IPOP 终端工具的输出状态，如图 6-8 和图 6-9 所示。

图 6-8　观察小车运行状态和输出状态

```
remote_control_car running
car_working() start
car_mode_control_func start
ConnectTo(0): 0
+NOTICE:SCANFINISH
+NOTICE:CONNECTED
netifapi_dhcp_start: 0
server :
        server_id : 192.168.0.1
        mask : 255.255.255.0, 0
        gw : 0.0.0.0
        T0 : 0
        T1 : 0
        T2 : 0
clients <1> :
        mac_idx mac              addr              state    lease    tries
        0       b4c9b9af689e     192.168.0.104     1        0        1
netifapi_netif_common: 0
mqtt client demo start
Sending to hostname 192.168.0.103 port 1883
message arrived 1,1s
remote_control_car_cmd:1
message arrived 1,0s
remote_control_car_cmd:0
message arrived 1,2s
remote_control_car_cmd:2
message arrived 1,0s
remote_control_car_cmd:0
```

图 6-9　小车端（IPOP 终端工具）的运行效果

注意：如果出现小车无反应现象，则应进行以下检测。

① 查看 IPOP 终端工具，确认小车是否成功联网。

② 确认小车和 MQTT 客户端工具是否与 MQTT 服务器建立连接。如果无法连接，则应开放服务器 1883 端口或关闭防火墙。

③ 确认小车订阅主题与 MQTT 客户端工具发布主题是否相同。

④ 确保发送的小车指令内容与小车指令枚举一致。

6.3　手表端功能实现

在 HarmonyOS 手表端创建 Java 模板遥控小车 App，实现 UI 控制小车指令的发布功能。实现 UI 设计和功能实现。

6.3.1　创建并配置工程

完成小车工程 WiFiCar 创建及工程配置，具体步骤如下：

（1）参考 HarmonyOS 官网或者《HarmonyOS 入门到精通 40 例》第 1 章完成 HarmonyOS 开发环境搭建。

（2）创建 Java 模板空工程 WiFiCar，将工程类型设置为 Application（应用）、将 SDK API 设置为 6、将设备类型设置为 Wearable（手表），如图 6-10 所示。

图 6-10 创建 WiFiCar 工程

(3) 将 App 的中文名称修改为"遥控小车"。具体操作是同时在 zh\element\string.json 和 base\element\string.json 文件中将 name 修改为 entry_MainAbility 并将 value 修改为"遥控小车",zh\element\string.json 文件中的代码如下:

```
//WiFiCar\entry\src\main\resources\zh\element\string.json
{
  "string": [
    {
      "name": "entry_MainAbility",
      "value": "遥控小车"
    },
    {
      "name": "mainability_description",
      "value": "Java_Empty Ability"
    },
    {
      "name": "mainability_HelloWorld",
      "value": "你好,世界"
    }
  ]
}
```

(4) base\element\string.json 文件的代码如下：

```json
//WiFiCar\entry\src\main\resources\base\element\string.json
{
  "string": [
    {
      "name": "entry_MainAbility",
      "value": "遥控小车"
    },
    {
      "name": "mainability_description",
      "value": "Java_Empty Ability"
    },
    {
      "name": "mainability_HelloWorld",
      "value": "你好,世界"
    }
  ]
}
```

(5) 将 App 的英文名称修改为 WiFiCar。具体操作是在 en\element\string.json 文件中将 name 修改为 entry_MainAbility 并将 value 修改为 WiFiCar，en\element\string.json 文件中的代码如下：

```json
//WiFiCar\entry\src\main\resources\en\element\string.json
{
  "string": [
    {
      "name": "entry_MainAbility",
      "value": "WiFiCar"
    },
    {
      "name": "mainability_description",
      "value": "Java_Empty Ability"
    },
    {
      "name": "mainability_HelloWorld",
      "value": "Hello World"
    }
  ]
}
```

(6) 修改 App 桌面快捷启动图标。具体操作是将像素为 114×114 的图片复制到 resources\base\media 文件夹中，在 config.json 文件中修改 ability 的 icon 属性，如图 6-11 所示。

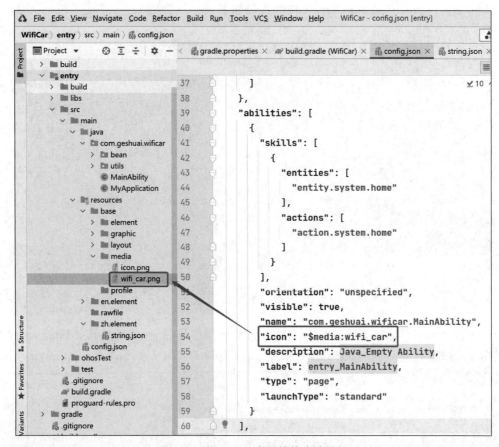

图 6-11 修改 App 桌面快捷启动图标

（7）在 config.json 文件中添加网络权限，config.json 文件中的完整代码如下：

```
//entry\src\main\config.json
{
  "app": {
    "bundleName": "com.geshuai.wificar",
    "vendor": "geshuai",
    "version": {
      "code": 1000000,
      "name": "1.0.0"
    }
  },
  "deviceConfig": {},
  "module": {
    "package": "com.geshuai.wificar",
    "name": ".MyApplication",
```

```json
      "mainAbility": "com.geshuai.wificar.MainAbility",
      "deviceType": [
        "wearable"
      ],
      "distro": {
        "deliveryWithInstall": true,
        "moduleName": "entry",
        "moduleType": "entry",
        "installationFree": false
      },
      "abilities": [
        {
          "skills": [
            {
              "entities": [
                "entity.system.home"
              ],
              "actions": [
                "action.system.home"
              ]
            }
          ],
          "orientation": "unspecified",
          "visible": true,
          "name": "com.geshuai.wificar.MainAbility",
          "icon": "$media:wifi_car",
          "description": "$string:mainability_description",
          "label": "$string:entry_MainAbility",
          "type": "page",
          "launchType": "standard"
        }
      ],
      "reqPermissions": [
        {
          "name": "ohos.permission.INTERNET"
        }
      ]
    }
}
```

6.3.2　UI 设计与实现

App 的 UI 用于实现 App 的名称、版本号、前进、后退、左转、右转、停止等 9 个按键的布局，具体操作如下：

（1）UI 布局设计，整体布局为垂直布局，总共分为 3 部分，从上到下依次为 App 名称、功能按键和App版本号,其中功能按键又分为上中下3行(3个水平布局),每行有3个按键,参考布局如图6-12所示。

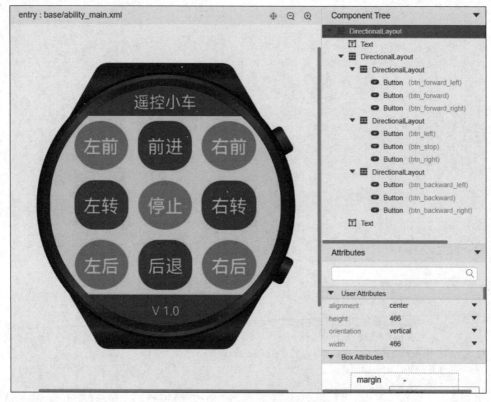

图6-12 遥控小车的UI布局

(2) 代码如下:

```
//entry\src\main\resources\base\layout\ability_main.xml
<?xml version = "1.0" encoding = "utf-8"?>
<DirectionalLayout
    xmlns:ohos = "http://schemas.huawei.com/res/ohos"
    ohos:height = "match_parent"
    ohos:width = "match_parent"
    ohos:alignment = "center"
    ohos:orientation = "vertical">
    <Text
        ohos:height = "match_parent"
        ohos:width = "match_parent"
        ohos:weight = "1"
        ohos:text = "遥控小车"
        ohos:text_size = "16fp"
```

```xml
        ohos:text_color = "white"
        ohos:text_alignment = "center"
        ohos:background_element = "blue"/>
<DirectionalLayout
    ohos:height = "match_parent"
    ohos:width = "match_parent"
    ohos:left_padding = "20vp"
    ohos:right_padding = "20vp"
    ohos:weight = "6"
    ohos:orientation = "vertical"
    >
    <DirectionalLayout
        ohos:height = "match_parent"
        ohos:width = "match_parent"
        ohos:orientation = "horizontal"
        ohos:weight = "1">
        <Button
            ohos:id = "$+id:btn_forward_left"
            ohos:height = "match_parent"
            ohos:width = "match_parent"
            ohos:align_parent_top = "true"
            ohos:background_element = "$graphic:background_button_oval"
            ohos:horizontal_center = "true"
            ohos:text = "左前"
            ohos:text_alignment = "center"
            ohos:margin = "5vp"
            ohos:weight = "1"
            ohos:text_color = "white"
            ohos:text_size = "18fp"
            />
        <Button
            ohos:id = "$+id:btn_forward"
            ohos:height = "match_parent"
            ohos:width = "match_parent"
            ohos:background_element = "$graphic:background_button_corner"
            ohos:horizontal_center = "true"
            ohos:text = "前进"
            ohos:text_alignment = "center"
            ohos:text_color = "white"
            ohos:margin = "5vp"
            ohos:weight = "1"
            ohos:text_size = "18fp"
            />
        <Button
            ohos:id = "$+id:btn_forward_right"
            ohos:height = "match_parent"
```

```xml
        ohos:width = "match_parent"
        ohos:align_parent_top = "true"
        ohos:background_element = " $ graphic:background_button_oval"
        ohos:horizontal_center = "true"
        ohos:text = "右前"
        ohos:text_alignment = "center"
        ohos:text_color = "white"
        ohos:margin = "5vp"
        ohos:weight = "1"
        ohos:text_size = "18fp"
        />
</DirectionalLayout>
< DirectionalLayout
    ohos:height = "match_parent"
    ohos:width = "match_parent"
    ohos:orientation = "horizontal"
    ohos:weight = "1">
    < Button
        ohos:id = " $ + id:btn_left"
        ohos:height = "match_parent"
        ohos:width = "match_parent"
        ohos:align_parent_left = "true"
        ohos:background_element = " $ graphic:background_button_corner"
        ohos:text = "左转"
        ohos:text_alignment = "center"
        ohos:text_color = "white"
        ohos:text_size = "18fp"
        ohos:margin = "5vp"
        ohos:weight = "1"
        ohos:vertical_center = "true"
        />
    < Button
        ohos:id = " $ + id:btn_stop"
        ohos:height = "match_parent"
        ohos:width = "match_parent"
        ohos:background_element = " $ graphic:background_button_oval"
        ohos:center_in_parent = "true"
        ohos:text = "停止"
        ohos:text_alignment = "center"
        ohos:text_color = "white"
        ohos:text_size = "18fp"
        ohos:weight = "1"
        ohos:margin = "5vp"
        />
    < Button
        ohos:id = " $ + id:btn_right"
```

```xml
        ohos:height = "match_parent"
        ohos:width = "match_parent"
        ohos:align_parent_right = "true"
        ohos:background_element = " $ graphic:background_button_corner"
        ohos:text = "右转"
        ohos:text_alignment = "center"
        ohos:text_color = "white"
        ohos:text_size = "18fp"
        ohos:margin = "5vp"
        ohos:weight = "1"
        ohos:vertical_center = "true"
        />
</DirectionalLayout>
< DirectionalLayout
    ohos:height = "match_parent"
    ohos:width = "match_parent"
    ohos:orientation = "horizontal"
    ohos:weight = "1">
    < Button
        ohos:id = " $ + id:btn_backward_left"
        ohos:height = "match_parent"
        ohos:width = "match_parent"
        ohos:align_parent_bottom = "true"
        ohos:background_element = " $ graphic:background_button_oval"
        ohos:horizontal_center = "true"
        ohos:text = "左后"
        ohos:text_alignment = "center"
        ohos:text_color = "white"
        ohos:margin = "5vp"
        ohos:weight = "1"
        ohos:text_size = "18fp"
        />
    < Button
        ohos:id = " $ + id:btn_backward"
        ohos:height = "match_parent"
        ohos:width = "match_parent"
        ohos:align_parent_bottom = "true"
        ohos:background_element = " $ graphic:background_button_corner"
        ohos:horizontal_center = "true"
        ohos:text = "后退"
        ohos:text_alignment = "center"
        ohos:text_color = "white"
        ohos:margin = "5vp"
        ohos:weight = "1"
        ohos:text_size = "18fp"
        />
```

```xml
            <Button
                ohos:id = " $ + id:btn_backward_right"
                ohos:height = "match_parent"
                ohos:width = "match_parent"
                ohos:align_parent_bottom = "true"
                ohos:background_element = " $ graphic:background_button_oval"
                ohos:horizontal_center = "true"
                ohos:text = "右后"
                ohos:text_alignment = "center"
                ohos:text_color = "white"
                ohos:margin = "5vp"
                ohos:weight = "1"
                ohos:text_size = "18fp"
                />
        </DirectionalLayout>
    </DirectionalLayout>
    <Text
        ohos:height = "match_parent"
        ohos:width = "match_parent"
        ohos:weight = "1"
        ohos:text = "V 1.0"
        ohos:text_size = "12fp"
        ohos:text_color = "white"
        ohos:text_alignment = "center"
        ohos:background_element = "blue"/>
</DirectionalLayout>
```

(3) 为按键前进、后退、左转、右转创建单击效果样式文件 background_button_corner.xml、默认状态样式文件 background_button_corner_normal.xml 和按下状态样式文件 background_button_corner_pressed.xml。

background_button_corner.xml 文件中的代码如下：

```xml
//RemoteControlCar\entry\src\main\resources\base\graphic\background_button_corner.xml
<?xml version = "1.0" encoding = "utf - 8"?>
<state - container xmlns:ohos = "http://schemas.huawei.com/res/ohos">
    <item ohos:state = "component_state_pressed" ohos:element = " $ graphic:background_button_corner_pressed"/>
    <item ohos:state = "component_state_empty" ohos:element = " $ graphic:background_button_corner_normal"/>
</state - container>
```

background_button_corner_normal.xml 文件中的代码如下：

```
//RemoteControlCar\ entry\ src\ main\ resources\ base\ graphic\ background_button_corner_normal.xml
```

```xml
<?xml version = "1.0" encoding = "UTF-8" ?>
<shape xmlns:ohos = "http://schemas.huawei.com/res/ohos"
       ohos:shape = "rectangle">
    <solid
        ohos:color = "#0000FF"/>
    <corners
        ohos:radius = "20vp"/>
</shape>
```

background_button_corner_pressed.xml 文件中的代码如下:

```xml
//RemoteControlCar\entry\src\main\resources\base\graphic\background_button_corner_pressed.xml
<?xml version = "1.0" encoding = "UTF-8" ?>
<shape xmlns:ohos = "http://schemas.huawei.com/res/ohos"
       ohos:shape = "rectangle">
    <solid
        ohos:color = "#FF0000"/>
    <corners
        ohos:radius = "20vp"/>
</shape>
```

（4）为按键左前、右前、左后、右后和停止创建单击效果样式文件 background_button_oval.xml、默认状态样式文件 background_button_oval_normal.xml 和按下状态样式文件 background_button_oval_pressed.xml。

background_button_oval.xml 文件中的代码如下:

```xml
//RemoteControlCar\entry\src\main\resources\base\graphic\background_button_oval.xml
<?xml version = "1.0" encoding = "utf-8"?>
<state-container xmlns:ohos = "http://schemas.huawei.com/res/ohos">
    <item ohos:state = "component_state_pressed" ohos:element = "$graphic:background_button_oval_pressed"/>
    <item ohos:state = "component_state_empty" ohos:element = "$graphic:background_button_oval_normal"/>
</state-container>
```

background_button_oval_normal.xml 文件中的代码如下:

```xml
//RemoteControlCar\entry\src\main\resources\base\graphic\background_button_oval_normal.xml
<?xml version = "1.0" encoding = "UTF-8" ?>
<shape xmlns:ohos = "http://schemas.huawei.com/res/ohos"
       ohos:shape = "oval">
    <solid
        ohos:color = "#FF9B9B9D"/>
</shape>
```

background_button_oval_pressed.xml 文件中的代码如下：

```xml
//RemoteControlCar \ entry \ src \ main \ resources \ base \ graphic \ background _ button _ oval _ pressed.xml
<?xml version = "1.0" encoding = "UTF - 8" ?>
< shape xmlns:ohos = "http://schemas.huawei.com/res/ohos"
       ohos:shape = "oval">
    <solid
        ohos:color = "#FF0000"/>
</shape>
```

6.3.3　功能实现

实现 MQTT 的客户端对象创建和按键响应功能。

（1）创建 bean 包，并在其中创建小车指令枚举类 CarStat，CarStat.java 文件中的代码如下：

```java
//entry\src\main\java\com\geshuai\wificar\bean\CarStat.java
package com.geshuai.wificar.bean;
public enum CarStat {
    /* 停止 */
    CAR_STATUS_STOP,
    /* 前进 */
    CAR_STATUS_FORWARD,
    /* 后退 */
    CAR_STATUS_BACKWARD,
    /* 左转 */
    CAR_STATUS_LEFT,
    /* 左前 */
    CAR_STATUS_LEFT_FRONT,
    /* 左退 */
    CAR_STATUS_LEFT_BACK,
    /* 右转 */
    CAR_STATUS_RIGHT,
    /* 右前 */
    CAR_STATUS_RIGHT_FRONT,
    /* 右退 */
    CAR_STATUS_RIGHT_BACK,
    /** Maximum value */
    CAR_STATUS_MAX
}
```

(2) 创建工具包 utils，在其中创建并编写工具类 ToastUtil.java，实现显示 Toast 功能，代码如下：

```java
//entry\src\main\java\com\geshuai\wificar\utils\ToastUtil.java
package com.geshuai.wificar.utils;
import ohos.agp.colors.RgbColor;
import ohos.agp.components.DirectionalLayout;
import ohos.agp.components.Text;
import ohos.agp.components.element.ShapeElement;
import ohos.agp.utils.Color;
import ohos.agp.utils.TextAlignment;
import ohos.agp.window.dialog.ToastDialog;
import ohos.app.Context;
public class ToastUtil {
    public static void show(Context context,String content,int duration){
        ToastDialog td = new ToastDialog(context);
        DirectionalLayout myLayout = new DirectionalLayout(context);
        td.setDuration(duration);
        DirectionalLayout.LayoutConfig config = new DirectionalLayout.LayoutConfig(DirectionalLayout.LayoutConfig.MATCH_CONTENT, DirectionalLayout.LayoutConfig.MATCH_CONTENT);
        myLayout.setLayoutConfig(config);
        ShapeElement element = new ShapeElement();
        element.setRgbColor(new RgbColor(255, 255, 255));
        myLayout.setBackground(element);
        Text text = new Text(context);
        text.setLayoutConfig(config);
        text.setText(content);
        text.setTextColor(new Color(0xFF000000));
        text.setTextSize(50);
        text.setTextAlignment(TextAlignment.CENTER);
        myLayout.addComponent(text);
        td.setComponent(myLayout);
        td.show();
    }
}
```

(3) 将 MQTT 的 SDK wmqtt.jar 文件复制到 libs 目录下，右击此文件后导入为库文件，如图 6-13 所示。

(4) 在 utils 目录下创建并编写工具类 MqttReceiverClient.java，代码如下：

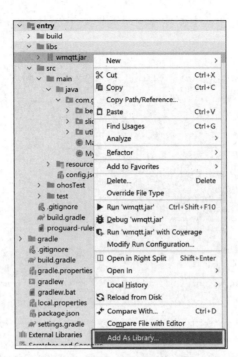

图 6-13 导入 MQTT 第三方库

```java
//entry\src\main\java\com\geshuai\wificar\utils\MqttReceiverClient.java
package com.geshuai.wificar.utils;
import com.ibm.mqtt.*;
import ohos.hiviewdfx.HiLog;
import ohos.hiviewdfx.HiLogLabel;
public class MqttReceiverClient {
    HiLogLabel hiLogLabel = new HiLogLabel(HiLog.LOG_APP, 0x10001, "mqtt");
    //定义 MQTT 连接状态的常量
    public enum MQTTConnectionStatus {
        INITIAL,                            //初始状态
        CONNECTING,                         //尝试连接
        CONNECTED,                          //已连接
        CONNECTEDANDSUB,                    //已连接并已订阅
        NOTCONNECTED_WAITINGFORINTERNET,    //无法连接,因为手机无法访问网络
        NOTCONNECTED_USERDISCONNECT,        //用户已明确请求断开连接
        NOTCONNECTED_DATADISABLED,          //无法连接,因为用户已禁用数据访问
        NOTCONNECTED_UNKNOWNREASON          //由于某种原因无法连接
    }
    //MQTT 常量
    public static final int MAX_MQTT_CLIENTID_LENGTH = 22;
    //MQTT 客户端连接的状态
    private MQTTConnectionStatus connectionStatus = MQTTConnectionStatus.INITIAL;
```

```java
private IMqttClient mqttClient = null;
private String brokerHostName = null;
private MqttPersistence usePersistence = null;
private boolean cleanStart = false;
private String mqttClientId = null;
private String[] topicsAll = {};
private int[] qualitiesOfService;
private short keepAliveSeconds = 20 * 60;
private boolean isFirstConnected = false;
MqttSimpleCallback myMqttSimpleCallback;

public MqttReceiverClient(MqttSimpleCallback myMqttSimpleCallback) {
    this.myMqttSimpleCallback = myMqttSimpleCallback;
}
public void onStart() {
    String mqttConnSpec = "tcp://192.168.0.103@1883";
    try {
        //定义到服务器的连接
        mqttClient = MqttClient.createMqttClient(mqttConnSpec,
            usePersistence);
        //注册此客户端应用程序,以便能够接收消息
        mqttClient.registerSimpleHandler(myMqttSimpleCallback);
    } catch (MqttException e) {
        //出了问题
        mqttClient = null;
        connectionStatus = MQTTConnectionStatus.NOTCONNECTED_UNKNOWNREASON;
    }
    connectToBroker();

}
private boolean isAlreadyConnected() {
    return ((mqttClient != null) && (mqttClient.isConnected() == true));
}
public boolean connectToBroker() {
    if (mqttClient == null)
        return false;
    boolean ret = false;
    try {
        try {
            if (isAlreadyConnected()) {
                mqttClient.unsubscribe(topicsAll);
                mqttClient.disconnect();
            }
        } catch (MqttPersistenceException ee) {
            HiLog.error(hiLogLabel, "mqtt", "disconnect failed - persistence exception",
ee);
```

```java
            } catch (MqttException ee) {
                ee.printStackTrace();
            }
            //尝试去连接
            mqttClient.connect("watch_gs", cleanStart, keepAliveSeconds);
            return true;
        } catch (MqttException e) {
            HiLog.error(hiLogLabel, "无法连接到推送服务器");
            HiLog.error(hiLogLabel, String.valueOf(e.getCause()));
            return ret;
        }
    }
    /**
     * 发布消息
     *
     * @param topic    主题
     * @param msg      消息
     * @param thisQoS  消息质量
     * @param retained 是否保留
     * @return
     */
    public void publish(final String topic, String msg, int thisQoS, boolean retained) {
        if (topic != null && !"".equals(topic) && msg != null && !"".equals(msg)) {
            try {
                mqttClient.publish(topic, msg.getBytes(), 0, false);
            } catch (Exception e) {
                e.printStackTrace();
            }
        }
    }
    public void publish(final String topic, String msg) {
        publish(topic, msg, 0, false);
    }
}
```

（5）在 MainAbilitySlice 中声明 MqttReceiverClient 类型的变量 mqttReceiverClient。

（6）在 onStart 方法中创建新线程，实现 MqttReceiverClient 的对象的创建及连接 MQTT 服务器的功能。

（7）MainAbilitySlice 类实现接口 Component.ClickedListener，重写 onClick 方法。

（8）通过 ID 将全部按键的单击监听器设置为当前 MainAbilitySlice 对象。

（9）在 onClick 方法中创建新线程，在线程中通过组件的 ID 判断不同的按键，发布对应的 MQTT 消息，MainAbilitySlice.java 文件中的代码如下：

```java
//entry\src\main\java\com\geshuai\wificar\slice\MainAbilitySlice.java
package com.geshuai.wificar.slice;
import com.geshuai.wificar.ResourceTable;
import com.geshuai.wificar.utils.MqttReceiverClient;
import com.ibm.mqtt.MqttSimpleCallback;
import ohos.aafwk.ability.AbilitySlice;
import ohos.aafwk.content.Intent;
import ohos.agp.components.Component;
import static com.geshuai.wificar.bean.CarStat.*;
public class MainAbilitySlice extends AbilitySlice implements Component.ClickedListener {
    MqttReceiverClient mqttReceiverClient;
    @Override
    public void onStart(Intent intent) {
        super.onStart(intent);
        super.setUIContent(ResourceTable.Layout_ability_main);
        new Thread(new Runnable() {
            @Override
            public void run() {
                mqttReceiverClient = new MqttReceiverClient(new MqttSimpleCallback() {
                    @Override
                    public void connectionLost() throws Exception {
                        int i = 0;
                        while (!mqttReceiverClient.connectToBroker() && i++< 5) {
                        }
                    }
                    @Override
                    public void publishArrived(String s, Byte[] Bytes, int i, boolean b) throws Exception {
                    }
                });
                mqttReceiverClient.onStart();
            }
        }).start();
        findComponentById(ResourceTable.Id_btn_forward_left).setClickedListener(this);
        findComponentById(ResourceTable.Id_btn_forward).setClickedListener(this);
        findComponentById(ResourceTable.Id_btn_forward_right).setClickedListener(this);
        findComponentById(ResourceTable.Id_btn_left).setClickedListener(this);
        findComponentById(ResourceTable.Id_btn_right).setClickedListener(this);
        findComponentById(ResourceTable.Id_btn_stop).setClickedListener(this);
        findComponentById(ResourceTable.Id_btn_backward_left).setClickedListener(this);
        findComponentById(ResourceTable.Id_btn_backward).setClickedListener(this);
        findComponentById(ResourceTable.Id_btn_backward_right).setClickedListener(this);
    }
    @Override
    public void onActive() {
        super.onActive();
```

```java
}
@Override
public void onForeground(Intent intent) {
    super.onForeground(intent);
}
@Override
public void onClick(Component component) {
    new Thread(new Runnable() {
        @Override
        public void run() {
            int stat;
            switch (component.getId()) {
                case ResourceTable.Id_btn_forward_left:
                    stat = CAR_STATUS_LEFT_FRONT.ordinal();
                    break;
                case ResourceTable.Id_btn_forward:
                    stat = CAR_STATUS_FORWARD.ordinal();
                    break;
                case ResourceTable.Id_btn_forward_right:
                    stat = CAR_STATUS_RIGHT_FRONT.ordinal();
                    break;
                case ResourceTable.Id_btn_left:
                    stat = CAR_STATUS_LEFT.ordinal();
                    break;
                case ResourceTable.Id_btn_right:
                    stat = CAR_STATUS_RIGHT.ordinal();
                    break;
                case ResourceTable.Id_btn_backward_left:
                    stat = CAR_STATUS_LEFT_BACK.ordinal();
                    break;
                case ResourceTable.Id_btn_backward:
                    stat = CAR_STATUS_BACKWARD.ordinal();
                    break;
                case ResourceTable.Id_btn_backward_right:
                    stat = CAR_STATUS_RIGHT_BACK.ordinal();        //访问请求
                    break;
                default:
                    stat = CAR_STATUS_STOP.ordinal();
                    break;
            }
            mqttReceiverClient.publish("res", stat + "");          //访问请求
        }
    }).start();
}
```

6.4 多端联调

进行 HarmonyOS 手表端、服务器端和 OpenHarmony 开发板端联调,测试通过 MQTT 协议实现手表端 App 操控 OpenHarmony 小车开发板。测试步骤如下:

(1) 参考 5.4.2 节,完成 MQTT 服务部署并启用,如图 6-14 所示。

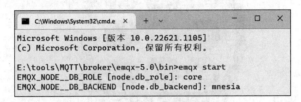

图 6-14　启动 MQTT 服务

(2) 运行 MQTT 客户端,连接 MQTT 服务器并订阅主题 res,用于接收和观察 HarmonyOS 端操作指令,如图 6-15 所示。

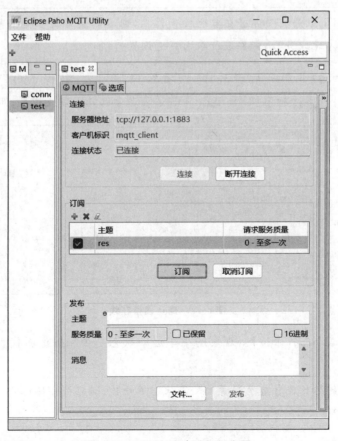

图 6-15　MQTT 客户端订阅主题

（3）运行 OpenHarmony 小车开发板端，确保 MQTT 服务连接成功。

（4）启动本地手表模拟器，在手表模拟器 WiFiCar 程序运行，如图 6-16 所示。

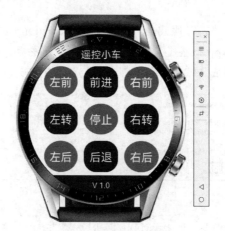

图 6-16　运行 WiFiCar 程序

注意：如果 MQTT 服务程序部署在本地服务器上，则不能使用远程模拟器，因为远程模拟器无法直接访问本地服务器，需要进行路由穿透设置。

（5）进入 MQTT 服务后台程序，查看客户端列表，确保 MQTT 客户端工具、OpenHarmony 小车开发板端和 HarmonyOS 手表端全部处于在线状态，如图 6-17 所示。

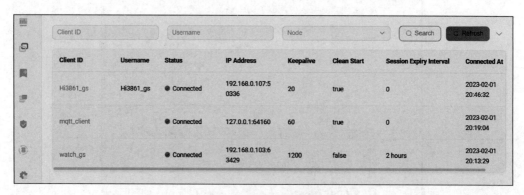

图 6-17　确认设备在线

（6）单击手表端的各个按键，观察的小车运动状态是否和按键所代表的状态一致，如图 6-18 所示。

注意：如果小车的运行状态与按键所代表的状态不一致，则应核实按键的 ID 与小车指令枚举，使其保持一致。

第6章 综合案例：遥控小车 271

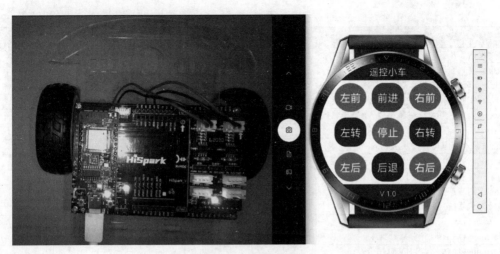

图 6-18 多端联调

图 书 推 荐

书 名	作 者
深度探索 Vue.js——原理剖析与实战应用	张云鹏
剑指大前端全栈工程师	贾志杰、史广、赵东彦
Flink 原理深入与编程实战——Scala+Java（微课视频版）	辛立伟
Spark 原理深入与编程实战（微课视频版）	辛立伟、张帆、张会娟
PySpark 原理深入与编程实战（微课视频版）	辛立伟、辛雨桐
HarmonyOS 移动应用开发（ArkTS 版）	刘安战、余雨萍、陈争艳 等
HarmonyOS 应用开发实战（JavaScript 版）	徐礼文
HarmonyOS 原子化服务卡片原理与实战	李洋
鸿蒙操作系统开发入门经典	徐礼文
鸿蒙应用程序开发	董昱
鸿蒙操作系统应用开发实践	陈美汝、郑森文、武延军、吴敬征
HarmonyOS 移动应用开发	刘安战、余雨萍、李勇军 等
HarmonyOS App 开发从 0 到 1	张诏添、李凯杰
HarmonyOS 从入门到精通 40 例	戈帅
JavaScript 基础语法详解	张旭乾
华为方舟编译器之美——基于开源代码的架构分析与实现	史宁宁
Android Runtime 源码解析	史宁宁
鲲鹏架构入门与实战	张磊
鲲鹏开发套件应用快速入门	张磊
华为 HCIA 路由与交换技术实战	江礼教
华为 HCIP 路由与交换技术实战	江礼教
openEuler 操作系统管理入门	陈争艳、刘安战、贾玉祥 等
恶意代码逆向分析基础详解	刘晓阳
深度探索 Go 语言——对象模型与 runtime 的原理、特性及应用	封幼林
深入理解 Go 语言	刘丹冰
Spring Boot 3.0 开发实战	李西明、陈立为
深度探索 Flutter——企业应用开发实战	赵龙
Flutter 组件精讲与实战	赵龙
Flutter 组件详解与实战	［加］王浩然（Bradley Wang）
Flutter 跨平台移动开发实战	董运成
Dart 语言实战——基于 Flutter 框架的程序开发（第 2 版）	亢少军
Dart 语言实战——基于 Angular 框架的 Web 开发	刘仕文
IntelliJ IDEA 软件开发与应用	乔国辉
Vue+Spring Boot 前后端分离开发实战	贾志杰
Vue.js 快速入门与深入实战	杨世文
Vue.js 企业开发实战	千锋教育高教产品研发部
Python 从入门到全栈开发	钱超
Python 全栈开发——基础入门	夏正东
Python 全栈开发——高阶编程	夏正东
Python 全栈开发——数据分析	夏正东
Python 编程与科学计算（微课视频版）	李志远、黄化人、姚明菊 等
Python 游戏编程项目开发实战	李志远
量子人工智能	金贤敏、胡俊杰
Python 人工智能——原理、实践及应用	杨博雄 主编，于营、肖衡、潘玉霞、高华玲、梁志勇 副主编
Python 预测分析与机器学习	王沁晨

续表

书 名	作 者
Python 数据分析实战——从 Excel 轻松入门 Pandas	曾贤志
Python 概率统计	李爽
Python 数据分析从 0 到 1	邓立文、俞心宇、牛瑶
FFmpeg 入门详解——音视频原理及应用	梅会东
FFmpeg 入门详解——SDK 二次开发与直播美颜原理及应用	梅会东
FFmpeg 入门详解——流媒体直播原理及应用	梅会东
FFmpeg 入门详解——命令行与音视频特效原理及应用	梅会东
Python Web 数据分析可视化——基于 Django 框架的开发实战	韩伟、赵盼
Python 玩转数学问题——轻松学习 NumPy、SciPy 和 Matplotlib	张骞
Pandas 通关实战	黄福星
深入浅出 Power Query M 语言	黄福星
深入浅出 DAX——Excel Power Pivot 和 Power BI 高效数据分析	黄福星
云原生开发实践	高尚衡
云计算管理配置与实战	杨昌家
虚拟化 KVM 极速入门	陈涛
虚拟化 KVM 进阶实践	陈涛
边缘计算	方娟、陆帅冰
物联网——嵌入式开发实战	连志安
动手学推荐系统——基于 PyTorch 的算法实现（微课视频版）	於方仁
人工智能算法——原理、技巧及应用	韩龙、张娜、汝洪芳
跟我一起学机器学习	王成、黄晓辉
深度强化学习理论与实践	龙强、章胜
自然语言处理——原理、方法与应用	王志立、雷鹏斌、吴宇凡
TensorFlow 计算机视觉原理与实战	欧阳鹏程、任浩然
计算机视觉——基于 OpenCV 与 TensorFlow 的深度学习方法	余海林、翟中华
深度学习——理论、方法与 PyTorch 实践	翟中华、孟翔宇
HuggingFace 自然语言处理详解——基于 BERT 中文模型的任务实战	李福林
Java＋OpenCV 高效入门	姚利民
AR Foundation 增强现实开发实战（ARKit 版）	汪祥春
AR Foundation 增强现实开发实战（ARCore 版）	汪祥春
ARKit 原生开发入门精粹——RealityKit ＋ Swift ＋ SwiftUI	汪祥春
HoloLens 2 开发入门精要——基于 Unity 和 MRTK	汪祥春
巧学易用单片机——从零基础入门到项目实战	王良升
Altium Designer 20 PCB 设计实战（视频微课版）	白军杰
Cadence 高速 PCB 设计——基于手机高阶板的案例分析与实战	李卫国、张彬、林超文
Octave 程序设计	于红博
Octave GUI 开发实战	于红博
ANSYS 19.0 实例详解	李大勇、周宝
ANSYS Workbench 结构有限元分析详解	汤晖
AutoCAD 2022 快速入门、进阶与精通	邵为龙
SolidWorks 2021 快速入门与深入实战	邵为龙
UG NX 1926 快速入门与深入实战	邵为龙
Autodesk Inventor 2022 快速入门与深入实战（微课视频版）	邵为龙
全栈 UI 自动化测试实战	胡胜强、单镜石、李睿
pytest 框架与自动化测试应用	房荔枝、梁丽丽